THE MATHESIS SERIES

Kenneth O. May, Editor

COUNTEREXAMPLES IN ANALYSIS

Bernard R. Gelbaum and John M. H. Olmsted

WAYS OF THOUGHT OF GREAT MATHEMATICIANS

Herbert Meschkowski

EVOLUTION OF MATHEMATICAL THOUGHT

Herbert Meschkowski

NUMBERS AND IDEALS

Abraham Robinson

ABRAHAM ROBINSON
University of California, Los Angeles

Numbers and Ideals

An introduction to some basic concepts
of algebra and number theory

HOLDEN-DAY, INC.
San Francisco, London, Amsterdam
1965

Library of Congress Catalog Card Number: 65–16747

Printed in the United States of America

Preface

This book has been written with a special purpose in mind, against the following background.

In the face of the vast expansion of available subject matter in mathematics, it is becoming increasingly difficult to design a suitable course of studies for undergraduates in this field. It seems highly desirable to introduce the student at the earliest possible moment to the modern methods which will enable him to appreciate, and later to take part in, contemporary research. Also, the power of these methods derives to a large extent from their great generality. But the beginner who is familiar only with a general theory and not with the concrete cases which are its roots and branches misses much of the beauty of the subject and may well be at a disadvantage when embarking on original research.

The present book attempts to bridge the gap between abstract methods and concrete applications by assembling its material around a central theme, the theory of ideals in algebraic (more particularly, quadratic) number fields. As it happens, the theory of ideals has turned out to be of basic importance not only in number theory but in various branches of abstract algebra and even in analysis. But some of the most fascinating aspects of the theory are present already in the concrete cases for which it was designed originally. It was the author's purpose, by focusing attention on such cases, to produce a concise text which is suitable for lower division and junior undergraduates, for high school teachers who wish to broaden their knowledge of mathematics, and even for gifted high school seniors.

Throughout, the emphasis is on *concepts*. The main purpose of the exercises is to serve as a test for the proper understanding of the subject matter.

Abraham Robinson

University of California
Los Angeles
April, 1965

Table of Contents

Introduction

Scientists, among them mathematicians, are driven by a lifelong desire to discover new worlds. Their reasons for doing so vary. They may have a practical purpose in mind. Or they may require some knowledge in a new field in order to solve an old problem. Or, like mountain climbers and explorers, they may want to tackle the unknown simply "because it is there." For a pure mathematician the world to be explored is in his own mind, although the field of application of his work may be in physics or technology. It follows that it is sometimes hard to say whether the mathematician invents a new subject, or discovers it.

The subject to which you will be introduced in this book was invented (or discovered) in order to solve a problem in pure mathematics which has no known application. Nevertheless, the notions and tools developed in order to explore this new world have in recent years proved useful even in physics and hence in technology. These applications are beyond our scope but we think that you will find the subject of sufficient interest in itself.

At the start, we shall make use of certain simple properties of the whole numbers, or *integers*, with which you are undoubtedly familiar, but we shall not discuss the question whether these properties can be reduced to even simpler ones. Later on you will on various occasions find a bracketed "Why?" in our text. This will indicate that we have intentionally left a small gap in our argument and that it is up to you to fill it.

I

The Ring of Integers and Some Other Rings

1. THE INTEGERS

The numbers 1, 2, 3, and so on, which are arrived at by the repeated addition of the number 1 (unity) to the number obtained previously—beginning with 1—are known as the positive whole numbers, or *positive integers*. If we add to this collection the number 0, then we obtain the *natural numbers*. 0 has the property that if we add it to any natural number, the result is the same number. Thus, $3 + 0 = 3$, $5 + 0 = 5$, $1 + 0 = 1$, $0 + 0 = 0$, or in general,

$$a + 0 = a\,.$$

Subtracting 1 repeatedly from the number obtained previously, beginning with 0, we obtain the *negative integers*, -1, -2, -3, and so on. Together with the natural numbers, the negative integers constitute the set of whole numbers or *integers*. An advantage of the set of all integers compared with the set of natural numbers alone is that within it we can always subtract one number from another. For example, $5 - 3 = 2$, $5 - 6 = -1$, $(-5) - (-7) = 2$, so that the result of subtracting an integer from another integer is an integer, where the result of subtracting one natural number from another may be a negative integer. A concrete way of representing the integers is obtained by marking off points at unit intervals (for example, one inch) on a straight line.

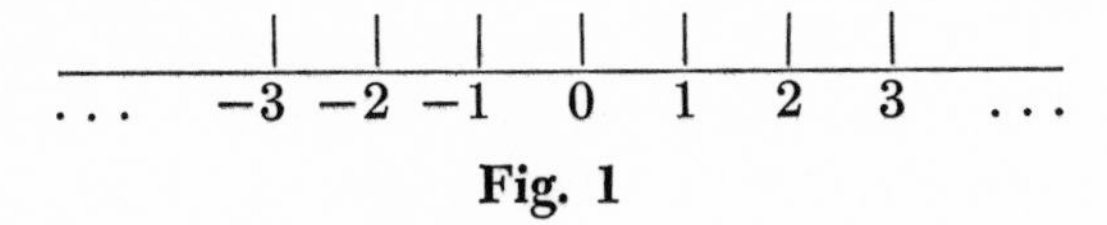

Fig. 1

The point corresponding to 0 is chosen arbitrarily (Fig. 1).

2. GROUPS

You may or may not have seen the following "laws of arithmetic" stated explicitly. In either case, you will recognize them as rules that you have used without hesitation in your computations with integers.

First Rule. When two integers are added together, the order in which they are taken does not affect the result. For example, $7 + 6 = 6 + 7 = 13$, $(-3) + 5 = 5 + (-3) = 2$, $4 + 0 = 0 + 4 = 4$, and in general

$$a + b = b + a\,,$$

where the letters a and b stand for arbitrary integers. This rule is called the *commutative law of addition.*

Second Rule. This is the *associative law of addition,*

$$(a + b) + c = a + (b + c)\,.$$

For example

$$(2 + 3) + 4 = 5 + 4 = 9$$

and

$$2 + (3 + 4) = 2 + 7 = 9$$

so that the two sums are equal.

Third Rule. There exists an integer, 0, whose addition to any integer yields that integer, in symbols,

$$a + 0 = a\,,$$

as mentioned earlier for the natural numbers. Because of this property, 0 is sometimes called the neutral element with respect to addition. There can be only one neutral element with respect to addition. (Why?)

Fourth Rule. To any integer a there exists another integer $b = -a$, such that

$$b + a = 0\,.$$

Thus, if a is a positive integer, then b is negative, if a is negative b is positive, and if $a = 0$ then $b = 0$. b is called the *inverse* of a.

It follows from these rules that for any two integers a and c, we can find just one integer x with the property that

$$x + a = c\,.$$

To see this (making use only of the first to fourth rule, and not of any other rule picked up elsewhere) let b be the inverse of a, $b = -a$, so that $b + a = 0$. Then the number $x = c + b = c + (-a)$ satisfies the equation $x + a = c$, since

$$x + a = (c + b) + a = c + (b + a) = c + 0 = c\,.$$

As usual, we write $x = c + (-a)$ more simply as $x = c - a$.

Fifth Rule. This rule states a fact which is so essential that until now we have taken it for granted. This is the fact that for any two integers, a and b, there exists a unique integer c such that c is the *sum* of a and b, $a + b = c$.

So far, we have enumerated a number of properties which are satisfied by the operation of addition within the set of all integers. It is natural to ask whether there are other sets of numbers (or even of some other kind of mathematical objects) which satisfy the rules just detailed. A simple example suggests itself as follows. Consider the set of all even integers, that is to say, the integers which are divisible (without remainder) by 2. The set includes 2, 4, 6, . . . but also 0, -2, -4, -6, and so on. All five rules are satisfied by the operation of addition within the domain of even integers. (Why?)

Any set in which an operation (here, the operation of addition) is given so that the five rules are satisfied is called a *commutative group*. The group is called *commutative* because of the first rule, which is the commutative law of addition. There exist groups which are not commutative but we shall not consider them in this book.

3. RINGS

The rules mentioned so far were concerned with addition.

Sixth Rule. For any two integers a and b there exists a unique integer c which is the *product* of a and b, $a \cdot b = c$ or $ab = c$, the result of multiplying a by b. The multiplication of integers has the following properties:

Seventh Rule. For any two integers a and b,

$$ab = ba\,.$$

This is the *commutative law of multiplication*. For example, $7 \cdot 8 = 56 = 8 \cdot 7$, $6 \cdot 0 = 0 = 0 \cdot 6$, $6 \cdot 1 = 6 = 1 \cdot 6$, $(-2) \cdot 3 = -6 = 3 \cdot (-2)$.

Eighth Rule. For any integers a, b, and c,

$$(ab)c = a(bc)\,.$$

This is the *associative law of multiplication.* For example, $(3 \cdot 7)8 = 21 \cdot 8 = 168$, $3(7 \cdot 8) = 3 \cdot 56 = 168$ and $[(-4) \cdot 2](-3) = (-8)(-3) = 24$, $(-4)[2 \cdot (-3)] = (-4)(-6) = 24$.

Ninth Rule. There exists an integer, 1, such that the multiplication of any integer by 1 yields that integer. In symbols

$$a \cdot 1 = a\,.$$

For example $(-5) \cdot 1 = -5$. Because of this property, 1 is called the neutral element of multiplication (or identity or *unit element*).

Tenth Rule. This is the *distributive law* which established a connection between addition and multiplication. It states that for any integers a, b, and c,

$$a(b + c) = ab + ac\,.$$

For example

$$7(3 + 4) = 7 \cdot 7 = 49$$

and

$$7 \cdot 3 + 7 \cdot 4 = 21 + 28 = 49\,.$$

Any system of numbers or of other mathematical objects which satisfies the ten rules enumerated so far is called a commutative ring with unit element. Here, the term "commutative" is due to the fact that the seventh rule is satisfied, and the term "unit element" refers to the number 1, which satisfies the ninth rule. Since we shall not consider any ring which does not satisfy the seventh rule we may as well omit the expression "commutative." It is easy to find a ring which does not satisfy the ninth rule. For example, the set of even numbers mentioned earlier satisfies the first to eighth, as well as the tenth rule. It does not include the number 1 (which is odd) and it does not include *any* number e with the property that $ae = a$ for all even numbers a.

Presently we shall give some more examples of rings, with or without unit elements. But first we remark that various rules of arithmetic which are familiar in the case of the integers (or of the real numbers) hold for all rings. For example

$$a \cdot 0 = 0$$

for all a in any given ring. To see this, we apply the tenth and third rules, so

$$a \cdot 0 = a\,(0 + 0) = a \cdot 0 + a \cdot 0\,.$$

Now, by the fourth rule, there exists an element b such that

$$b + a \cdot 0 = 0\,.$$

Hence, since we have already shown that

$$a \cdot 0 + a \cdot 0 = a \cdot 0\,,$$

it follows that

$$b + (a \cdot 0 + a \cdot 0) = b + a \cdot 0 = 0\,,$$

or

$$(b + a \cdot 0) + a \cdot 0 = 0 + a \cdot 0 = 0,$$

from which

$$a \cdot 0 = 0\,.$$

Another rule which is true in all rings is

$$a(-b) = -ab\,.$$

That is to say, the product of a and the inverse of b is equal to the inverse of ab. (Why?)

You are of course familiar with the custom of denoting numbers by letters and we have done so many times already. However, we find it convenient to denote entire sets of numbers or of other mathematical objects by letters, usually capital letters. Accordingly we agree to denote the set of positive integers by M, and the set of natural numbers by N (so that N differs from M only through the inclusion of 0 in N). The ring of integers will be denoted by J. Also, for any natural number n greater than 1, J_n shall denote the set of integers which are divisible (without remainder) by n. For example, J_2 is the set of even integers considered earlier, J_3 is the set of all integers which are multiples of 3, that is, 3, 6, 9, . . . and also 0, −3, −6, −9, ... All these J_n are rings without unit.

We shall now give a set of examples of rings with unit elements. Let n be any positive integer, for instance $n = 5$. We introduce a ring R_n which consists of the n numbers $0, 1, 2, \ldots, n-1$. The two operations of addition and multiplication are defined as in J except that if the result of the operation is greater than $n - 1$, then we take the remainder with respect to n. In order to distinguish these operations from ordinary addition and multiplication we shall write them in the form $a \oplus b$ and $a \odot b$. For example for $n = 5$,

$$2 + 2 = 2 \oplus 2 = 4$$

but

$$2 + 4 = 6$$

while $2 \oplus 4$ is the remainder of 6 after division by 5 and so

$$2 \oplus 4 = 1 .$$

Similarly

$$2 \cdot 2 = 2 \cdot 2 = 4 ,$$

but

$$2 \cdot 4 = 8 \text{ and } 2 \odot 4 = 3 .$$

Figures 2 and 3 are the complete tables of addition and multiplication for $n = 5$. The rows are numbered 0 to 4, corresponding to a, and the columns are similarly numbered, corresponding to b. At the intersection of the row labelled a and the column labelled b, Figure 2 then gives the *sums* $a \oplus b$ and Figure 3, the *product* $a \odot b$.

		b				
	$a \oplus b$	0	1	2	3	4
	0	0	1	2	3	4
	1	1	2	3	4	0
a	2	2	3	4	0	1
	3	3	4	0	1	2
	4	4	0	1	2	3

Fig. 2

		b				
	$a \odot b$	0	1	2	3	4
	0	0	0	0	0	0
	1	0	1	2	3	4
a	2	0	2	4	1	3
	3	0	3	1	4	2
	4	0	4	3	2	1

Fig. 3

You may now check that the ten rules which define a ring are all satisfied. For example, $2 \oplus 4 = 1 = 4 \oplus 2$, in agreement with the commutative law of addition, $4 \odot 3 = 3 \odot 4 = 2$ in agreement with the commutative law of multiplication. The number 0 is the neutral element with respect to addition, and every number has an inverse, in agreement with the fourth rule. Indeed

$$0 \oplus 0 = 0\,, \quad 4 \oplus 1 = 0\,, \quad 3 \oplus 2 = 0\,, \quad 2 \oplus 3 = 0\,, \quad 1 \oplus 4 = 0\,,$$

so that the inverses of 0, 1, 2, 3, 4 are 0, 4, 3, 2, 1 in that order.

Even if you have checked all possible cases for the ten rules you will only have proved the fact that our procedure leads to a ring for the particular case $n = 5$. However, we shall show later that this is actually true for all positive integers n.

In Figures 4 and 5 you will find the tables of addition and multiplication for the case $n = 6$ and, here again, you may check that the result is a ring.

	$a \oplus b$	0	1	2	3	4	5
	0	0	1	2	3	4	5
	1	1	2	3	4	5	0
	2	2	3	4	5	0	1
a	3	3	4	5	0	1	2
	4	4	5	0	1	2	3
	5	5	0	1	2	3	4

(columns: b)

Fig. 4

	$a \odot b$	0	1	2	3	4	5
	0	0	0	0	0	0	0
	1	0	1	2	3	4	5
	2	0	2	4	0	2	4
a	3	0	3	0	3	0	3
	4	0	4	2	0	4	2
	5	0	5	4	3	2	1

(columns: b)

Fig. 5

Comparing the multiplication table for this case (Fig. 5) with the multiplication table for $n = 5$ (Fig. 3), we notice an interesting difference. In Figure 3 all the zeros appear either in the row corresponding to $a = 0$ or in the column corresponding to $b = 0$. That is to say, a product is zero only if one of its two factors is zero. This is a property which is equally true for the integers, and we lay it down explicitly as our

Eleventh Rule. A product $a \cdot b$ equals zero only if at least one of the two factors is equal to zero.

An important consequence of this rule is the *cancellation rule.* If $ac = bc$ and $c \neq 0$, then $a = b$. For if $ac = bc$, then $0 = ac - bc = (a - b)c$ and if $c \neq 0$ then $a - b = 0$ by the eleventh rule, so that $a = b$.

Looking at Figure 5, on the other hand, we find that there are four zeros which are neither in the row $a = 0$ nor in the column $b = 0$. In other words, certain products are zero although none of their factors are. These are the products $2 \odot 3$, $3 \odot 2$, $3 \odot 4$, $4 \odot 3$. Any ring with unit element 1 such that $1 \neq 0$ and which satisfies the eleventh rule is called an *integral domain*. We see therefore that for $n = 5$ we have obtained an integral domain, while for $n = 6$ we have obtained a ring with unit element which is not an integral domain.

Disregarding the fact that we have not yet proved that the procedure just described leads to rings for all n, we shall call the resulting systems remainder rings or, to use a technical term, *residue rings*, and we shall denote them by R_n. ("Residue" is merely a more technical word for "remainder.")

4. ORDER

We are now going to discuss a property of the integers which is not shared by all other rings. This is the property of being *ordered* so as to satisfy certain "natural" rules. Thus we distinguish between greater and smaller integers, an integer a being smaller than an integer b, $a < b$, if the difference $b - a$ is a positive integer. Indeed, this may be regarded as the definition of the relation of order within the ring of integers. The fundamental properties of this relation are the following *rules of order:*

(i) For any integers a and b, precisely one of the following three relations holds: $a = b$, or $a < b$, or $b < a$.

(ii) If $a < b$ and $b < c$ then $a < c$.

This is the *rule of transitivity.*

(iii) If $a < b$ then $a + c < b + c$, for any integers a, b, and c.

For example, since $5 < 10$, we may conclude $5 + 3 < 10 + 3$ and $5 + (-8) < 10 + (-8)$, in other words $5 - 8 < 10 - 8$, in other words, $-3 < 2$.

(iv) If $a < b$ and $0 < c$ then $ac < bc$.

For example, since $5 < 10$ and $0 < 3$, we may conclude $5 \cdot 3 < 10 \cdot 3$, that is $15 < 30$. The condition $0 < c$ is essential. Thus, if we take $c = -1$, and try to draw the conclusion $ac < bc$ from $a < b$ for $a = 5$, $b = 10$, we obtain $-5 < -10$, which is wrong.

We shall sometimes write $a > b$ in place of $b < a$, and $a \leq b$ if either $a < b$ or $a = b$, and $a \geq b$ if either $a > b$ or $a = b$. We shall also take some of the simplest properties of the relation of order for granted.

5. INDUCTION

Completing our list of fundamental rules satisfied by natural numbers or by integers, we finally come to the *rule of induction.* We consider properties which are applicable to (that is, meaningful for) the natural numbers, such as the property of being an even number, or the property of ending with the digit 3 when written in decimal notation. Another property which is applicable to the natural numbers is that of being both even and odd. This property is meaningful for the natural numbers even though it is possessed by no natural number at all. Similarly, the property of being either even or odd is meaningful in relation to the natural numbers and is in fact possessed by all of them. On the other hand, the property of being written in blue ink does not apply to the natural numbers, but to each particular occurrence of a symbol which denotes a natural number. Such properties will not be considered here.

Every property which is applicable to the natural numbers determines a set, or collection, of natural numbers; that is, the collection, or set, of natural numbers which possesses that property. For example, the set of natural numbers that is determined by the property of being odd is composed of 1, 3, 5, 7, 9, 11, ... Since this set is infinite we cannot write all its elements down, and we have to rely on the reader to interpret the three dots after the number 11 correctly. Given any property P, it is convenient to denote the set of numbers having this property by the symbol

$$\{x|x \text{ possesses the property } P\}$$

for example

$$\{x|x \text{ is odd}\}$$

(read "the set of all x such that x is odd"). The same notation is used in other cases; for example, the property of revolving around the sun determines the set

$$\{x|x \text{ revolves around the sun}\}$$

which includes the planets, the planetoids, the comets, and, at time of writing, several man-made objects.

Coming back to the natural numbers, consider the property Q of having 1, 3, 5, 7, or 9 in the last digit when written in decimal notation. The set determined by this property is to be written as

$$\{x| \text{ The decimal representation of } x \text{ ends on one of the digits } 1, 3, 5, 7, 9\}.$$

But this is again the set of odd natural numbers, even though some knowledge of arithmetic is required in order to appreciate this. As it happens, we shall be interested in the sets of natural numbers with certain properties rather than in the properties as such.

Rule of Induction. Let P be a property which is applicable to the natural numbers. Suppose that 0 has the property P. Suppose in addition that whenever a natural number n has the property P, the natural number $n + 1$ also has the property P. Then all natural numbers have the property P.

As a simple application of the rule of induction let us prove that in any array of three consecutive natural numbers there is one at least which is divisible by 3.

Proof. Any three consecutive natural numbers can be written as n, $n + 1$, $n + 2$, where n is the smallest among the three numbers. A natural number n is divisible by 3 if $n = 3k$ for some natural number k. We have to show that for all natural numbers n, one at least of the three numbers n, $n + 1$, $n + 2$, is equal to $3k$ for some natural number k.

Now this assertion is certainly true for $n = 0$. For in this case, the three consecutive numbers are 0, 1, 2; and 0 is divisible by 3 since $0 = 3 \cdot 0$. Suppose that we have proved the theorem already for some particular n, then we wish to show that it is true also for $n + 1$. In other words, we suppose that there is at least one number divisible by 3 (or as we say also, at least one *multiple* of 3) among n, $n + 1$, $n + 2$, and we wish to show that there is at least one multiple of 3 among $n + 1$, $n + 2$, $n + 3$. Now if n is divisible by 3, $n = 3k$, then $n + 3 = 3k + 3 = 3(k + 1)$ so that the last number in the second triple is divisible by 3. If n is not divisible by 3, then by assumption either $n + 1$ or $n + 2$ must be divisible by 3 since one at least of n, $n + 1$, $n + 2$ is divisible by 3. Both $n + 1$ and $n + 2$ belong also to the second triple, so that again at least one number of that triple is divisible by 3.

Let P be the following property, which is at any rate meaningful for (or as we called it earlier, applicable to) all natural numbers n.

Property P: n is the smallest number n in a triple of consecutive numbers—that is, $n, n + 1, n + 2$—of which one at least is divisible by 3.

Then we have shown that the number 0 has this property; and that if a natural number n has the property P, then $n + 1$ also has the property P. Accordingly, the rule of induction proves that all natural numbers have the property P. This confirms our assertion.

We have *not* proved that a triple of consecutive natural numbers contains *no more* than one number which is divisible by 3. This may appear obvious and, at any rate can be proved even more easily. (Why?)

6. PRINCIPLE OF THE SMALLEST NUMBER

We shall now establish a very general rule which is a consequence of the rule of induction (given some simple properties of the natural numbers in addition to that rule).

Principle of the Smallest Number. Let Q be a set, or collection, of natural numbers. Suppose that Q is not empty. That is to say that there exists at least one natural number which belongs to Q. Then Q contains a smallest element, q. That is to say, there exists a natural number q which belongs to Q such that no natural number which is smaller than q belongs to Q.

This principle is still true if Q includes *all* natural numbers. In that case, q is simply the number 0.

To prove the principle of the smallest number, suppose that there exists a non-empty set of natural numbers Q for which the principle is false. Consider the following property P, which is applicable to all natural numbers n.

Property P: n is a natural number such that no natural number m which is smaller than n ($m < n$) belongs to Q.

Then 0 has the property P; there are no natural numbers smaller than 0 which belong to Q, since there are no natural numbers at all which are smaller than 0. Suppose that the natural number n has the property P. Then the natural numbers which are smaller than n do not belong to Q. If n belongs to Q then n is the smallest number which belongs to Q, contrary to the assumption that the principle of the smallest number is false for the set Q. Hence, n does not belong to Q. Hence, no number which is smaller than $n + 1$ belongs to Q. Hence, $n + 1$ has the property P. Hence, by the rule of induction, all natural numbers have the property P. But this would imply that there does not exist any natural number which belongs to Q. This is contrary to the assumption that Q is not empty. Accordingly we see that the assumption that the set Q does not contain a smallest number leads to a contradiction. We conclude that Q must contain a smallest number in all cases. This proves the principle of the smallest number.

We may also derive the rule of induction from the principle of the smallest number (given some simple properties of the natural numbers in addition to that principle). Suppose that P is a property which is applicable to the natural numbers, such that 0 has the property P and such that whenever a natural number n has the property P, $n + 1$ also has that property. Let Q be the set of natural numbers which do not have the property P. If Q is empty then the rule of induction is proved for this case. If Q is not empty then, by the principle of the smallest number, Q contains a smallest number, q. Now q cannot be 0, for 0 has the property P, by assumption. If q is greater than 0 then $n = q - 1$ is a natural number which does not belong to Q (since q is the smallest number of that description). But if so, then n has the property P and so $n + 1 = (q - 1) + 1 = q$ also has the property P, by assumption. It follows that q does not belong to Q, contrary to the definition of q. Thus the assumption that Q is not empty leads to a contradiction and the rule of induction is proved.

It is sometimes natural to use a *modified rule of induction* as follows.

Let P be a property which is applicable to (meaningful for) all natural numbers greater than or equal to some natural number n_0 which is assigned in advance. Suppose that n_0 has the property P. Suppose also that whenever a natural number $n \geq n_0$ has the property P, $n + 1$ also has that property. Then all natural numbers greater than or equal to n_0 have the property P.

This modified rule of induction follows without difficulty from our original rule. (Why?) In particular, if $n_0 = 1$, the rule applies to all positive integers.

EXERCISES

1. Which of the following is a group?

i. The rational numbers for the operation of multiplication
ii. The positive rational numbers for addition
iii. The positive rational numbers for multiplication.

2. Let A be a set of integers which is a group with respect to the operation of addition. Prove that there exists a natural number n such that A consists of all numbers that are divisible by n.

3. Which of the following is a ring?

i. The even integers
ii. The numbers $a + b\sqrt{2}$ where a and b are rational
iii. The odd integers.

4. Prove by induction that

$1 + 3 + 5 + \ldots + (2n + 1) = (n + 1)^2$ for $n = 0, 1, 2, 3, \ldots$.

5. Prove by induction that the number of ways in which n objects can be arranged in a row is

$$1 \cdot 2 \cdot 3 \cdot 4 \cdot 5 \cdot \ldots \cdot n = n!$$

II
Prime Numbers and Composite Numbers

1. PRIMES

Let n be any positive integer. n is divisible by itself (since $n = n \cdot 1$) and n is divisible by 1 (since $n = 1 \cdot n$). If n is greater than 1 and is divisible by no natural number other than itself or 1, then n is called *prime* or *a prime*. For example, 6 is not a prime since it is divisible by 2 and by 3. The numbers 2, 3, 5, and 7 are prime. You will have no difficulty in discovering further primes.

A positive integer greater than 1 which is not prime is called *composite*. If a positive integer n is divisible by a positive integer k then k is called a *divisor* of n. Thus, a prime number is a positive integer p greater than 1 whose only divisors are 1 and p. A divisor k of a positive integer n cannot be greater than n. (Why?)

Any positive integer greater than 1 must have at least one prime divisor. To see this, suppose that it is not true. Then the set Q of positive integers greater than 1 which do not have any prime divisors is not empty. According to the principle of the smallest number, Q contains a smallest number q. But q cannot be prime for then q would have itself as a prime divisor. Hence q is composite $q = mk$ where k is a positive integer smaller than q but greater than 1. Since k is smaller than q it does not belong to Q and so has a prime divisor; call it p. Then $k = rp$, and so $q = mk = (mr)p$. This shows that p is also a prime divisor of q. But q does not have any prime divisors. Accordingly, the assumption that there are positive integers greater than 1 which do not have any prime divisors leads to a contradiction. This shows that all positive integers greater than 1 have prime divisors.

2. EUCLID'S THEOREM

How many primes are there? The answer is given by a famous theorem which is credited to the Greek mathematician Euclid.

Theorem. The number of primes is infinite.

Proof. Let n be any positive integer, and let $n!$ be the factorial of n; that is, the product of all positive integers up to n. Thus

$$1! = 1, \quad 2! = 1 \cdot 2 = 2, \quad 3! = 1 \cdot 2 \cdot 3 = 6, \text{ and so on.}$$

We shall show that there is at least one prime number between any positive integer n and the larger number $m = n! + 1$, excluding n but including m. We may test this assertion for some values of n. For $n = 3$, $m = 7$ so that the range of numbers 4, 5, 6, 7 contains not one but two primes. For $n = 4$, $m = 25$, and the range in question again contains a relatively large number of primes: 5, 7, 11, 13, 17, 19, 23. This suggests that on the whole we may expect to find rather more than one prime number between n and $n! + 1$, and this is indeed true, though it is more difficult to prove than the simple assertion that there is at least one prime between n and $m = n! + 1$.

Consider a particular positive integer n. If $m = n! + 1$ is a prime then we have proved the theorem for this case. If m is not a prime then, as shown, there exists a prime k, neither 1 nor m, which divides m. k must be smaller than m. Now suppose that k is one of the numbers 1, 2, . . . , n. Then $m = pk$, for some positive integer p, since k divides m, and $n! = qk$ for some positive integer q, since k appears as a factor in $n!$ and hence, divides $n!$. We therefore have the two equations

$$\begin{aligned} n! + 1 &= pk \\ n! &= qk\,. \end{aligned}$$

Subtracting the second equation from the first, we obtain

$$1 = pk - qk = (p - q)k\,,$$

and so k divides 1. But the only positive integer that divides 1 is 1 itself (Why?) and k is different from 1. We have therefore arrived at a contradiction and conclude that k is one of the numbers $n + 1$, $n +$ 2, . . . , $n!$.

The example $n = 4$, $m = 25$ shows that it is quite possible that m is not a prime. But, in agreement with our proof, m is divisible by a prime k greater than n, that is, $k = 5$.

We now see that the number of primes is infinite. For if not, there would be a last (greatest) prime p. (Why?) But we have just shown that for any prime p there exists another prime greater than p (and not greater than $p! + 1$). Hence no p can be the last prime. We conclude that the number of primes is infinite.

3. THE EUCLIDEAN ALGORITHM

Let a and b be two positive integers such that $a \geq b$. To divide a by b with remainder means to find natural numbers q and r such that

$$a = qb + r \text{ where } 0 \leq r \leq b - 1 . \tag{3.1}$$

For example, if $a = 100$ and $b = 9$ then

$$100 = 11 \cdot 9 + 1$$

so that suitable values for q and r are $q = 11$, $r = 1$.

In order to prove that numbers q and r as specified can always be found, given a and b, we proceed by induction, using the *modified* rule (see page 12). Thus, let P be the following property of natural number n.

Property P: $n = a$ is positive, and for every positive integer b such that $a \geq b$ there exist natural numbers q and r which satisfy (3.1).

If $a = n = 1$ then $a \geq b$ implies $b = 1$ so that (3.1) is satisfied by $q = 1$, $r = 0$. Thus, P is possessed by $n = 1$. Suppose $n \geq 1$ possesses the property P, and let $a = n + 1$ and $b \leq a$. If $b = a$ then (3.1) is again satisfied by $q = 1$, $r = 0$. If $b < a$, then $b \leq n$ and so, by the assumption of induction, there exist natural numbers q' and r' such that

$$n = q'b + r' \quad 0 \leq r' \leq b - 1$$

and hence

$$a = n + 1 = q'b + (r' + 1) .$$

This shows that (3.1) is satisfied by $q = q'$, $r = r' + 1$, provided $r' < b - 1$. If $r' = b - 1$ then

$$a = n + 1 = q'b + b = (q' + 1)b$$

so that (3.1) is satisfied by $q = q' + 1$, $r = 0$. This completes the proof of the assertion that *division with remainder* is always possible.

We now introduce a chain, or sequence, of formulae of the type of (3.1) as follows. For given positive integers a_0 and a_1 such that $a_0 \geq a_1$ we write

$$a_0 = q_0a_1 + a_2 \text{ where } 0 \leq a_2 < a_1$$

for certain natural numbers q_0 and a_2. This is possible by (3.1). If $a_2 \neq 0$, then we write down the corresponding formula for a_1 and a_2, that is

$$a_1 = q_1 a_2 + a_3 \quad 0 \leq a_3 < a_2 ,$$

and we continue in this way until we come to a point where $a_{n+2} = 0$, that is, the remainder vanishes for the first time,

$$\begin{aligned} a_n &= q_n a_{n+1} + a_{n+2} \\ &= q_n a_{n+1} . \end{aligned}$$

Until this happens, we have $a_0 \geq a_1 > a_2 > a_3 > \ . \ . \ . > a_{n+1}$, a decreasing sequence of positive integers. Evidently, there can be no more than a_0 elements in this chain.

For example, if $a_0 = 72$ and $a_1 = 20$ then

$$\begin{aligned} 72 &= 3 \cdot 20 + 12 \\ 20 &= 1 \cdot 12 + 8 \\ 12 &= 1 \cdot 8 + 4 \\ 8 &= 2 \cdot 4 \end{aligned}$$

so that $a_2 = 12$, $a_3 = 8$, $a_4 = 4$, $a_5 = 0$.

We observe that for given a and b, $a \geq b > 0$, the natural numbers q and r are determined uniquely by (3.1). For suppose that

$$a = qb + r \quad 0 \leq r \leq b - 1 \tag{3.2}$$

and

$$a = q'b + r' \quad 0 \leq r' \leq b - 1 . \tag{3.3}$$

Then if $r = r'$, $qb = q'b$ and so $(q - q')b = 0$ and hence $q = q'$, as required. We are going to show that $r \neq r'$ is impossible. For if $r \neq r'$ then we may suppose without loss of generality that $r > r'$. (Why?) Subtracting (3.3) from (3.2), we obtain

$$0 = a - a = (qb + r) - (q'b + r') = (q - q')b + (r - r')$$

so that

$$(q' - q)b = r - r' .$$

But $0 < r - r' \leq r \leq b - 1$ so that the right-hand side of this equation (and hence its left-hand side) is positive but smaller than b. But if so then $q' - q$ must be a positive integer and so $(q' - q)b \geq b$, which contradicts $r - r' \leq b - 1$. This proves the uniqueness of q and r.

It follows that for given a_0, a_1, as above, the sequence a_0, a_1, a_2,

$\ldots, a_{n+1}, a_{n+2} = 0$ is determined uniquely. The computational procedure which leads up to it is known as the *Euclidean algorithm.*

Suppose now that a positive integer d divides the last remainder which is different from 0, a_{n+1}. Thus, $a_{n+1} = b_{n+1}d$ where b_{n+1} is some natural number. Since $a_n = q_n a_{n+1}$, it follows that $a_n = (q_n b_{n+1})d = b_n d$, where $b_n = q_n b_{n+1}$. Thus d divides also a_n. The preceding line in the algorithm (which we have not yet written down explicitly) is

$$a_{n-1} = q_{n-1}a_n + a_{n+1} = (q_{n-1}b_n)d + b_{n+1}d = (q_{n-1}b_n + b_{n+1})d\,.$$

Thus, d divides also a_{n-1}. Going back step by step, we conclude that d divides $a_{n+1}, a_n, a_{n-1}, a_{n-2}, \ldots$, so that finally, we find that d divides also a_1 and a_0. For example in the numerical case calculated above, 2 divides $a_4 = 4$ and hence, divides also $a_3 = 8$, $a_2 = 12$, $a_1 = 20$ and $a_0 = 72$.

Now let d' be any number which divides a_0 and a_1. We write the equations of the Euclidean algorithm in the following way

$$\begin{aligned} a_2 &= a_0 - q_0 a_1 \\ a_3 &= a_1 - q_1 a_2 \\ &\cdot \\ &\cdot \\ &\cdot \\ a_n &= a_{n-2} - q_{n-2}a_{n-1} \\ a_{n+1} &= a_{n-1} - q_{n-1}a_n \end{aligned} \tag{3.4}$$

(where the equation $a_n = q_n a_{n+1}$ has been disregarded). The first of these equations shows that d' divides also a_2. (Why?) The second equation shows that, since d' divides a_2 and a_1 it divides also a_3. Continuing in this way, we find that d' divides also $a_4, a_5, \ldots, a_{n+1}$.

4. THE GREATEST COMMON DIVISOR

We now introduce the notion of the *greatest common divisor* (G.C.D.) of the positive integers a_0 and a_1. This is a positive integer d^* which possesses the following properties:

(i) d^* divides both a_0 and a_1;
(ii) If d' is any positive integer which divides both a_0 and a_1, then d' divides also d^*.

We may ask whether it is possible that the positive integers a_0 and a_1 possess two different greatest common divisors. In a way we have already anticipated the answer to this question by talking of *the* greatest common divisor. However, suppose that there are two positive integers which possess the properties of d^* according to (i) and (ii), d_1^* and d_2^*. Then by (ii), with $d' = d_1^*$, $d^* = d_2^*$; d_1^* divides d_2^*. Interchanging d_1^* and d_2^* and applying (ii) again, we find that d_2^* divides d_1^*. Hence there exist positive integers e_1 and e_2 such that $d_2^* = e_1 d_1^*$ and $d_1^* = e_2 d_2^* = e_2 e_1 d_1^*$. Hence $e_2 e_1 = 1$, and so $e_2 = e_1 = 1$ (since the only positive integer which divides 1 is 1 itself). Hence $d_2^* = d_1^*$, d^* is unique in the sense that a_0 and a_1 cannot have *more* than one greatest common divisor. To show that a_0 and a_1 must have *at least* one greatest common divisor we distinguish two cases. If $a_0 = a_1$, then $d^* = a_0 = a_1$ is the greatest common divisor of a_0 and a_1. (Why?) If they are unequal, suppose $a_0 > a_1$ and carry out the Euclidean algorithm. As shown, the last non-zero remainder, a_{n+1}, divides a_0 and a_1 (since it divides itself). Also, by what we have shown already, any positive integer which divides a_0 and a_1 divides also a_{n+1}. It follows that $d^* = a_{n+1}$ satisfies the conditions (i) and (ii) and is the G.C.D. of a_0 and a_1.

We write $d = <a,b>$ if d is the greatest common divisor of the numbers a and b. For example, $<72, 20> = 4$.

If an integer d divides integers a and b then it divides also all integers which can be written as $ha + kb$ where h and k are any integers. For example, since 5 divides $a = 10$ and $b = 15$ it divides also $7 \cdot 10 - 3 \cdot 15 = 25$. We are going to show that for any positive integers a_0, a_1, their G.C.D. $d^* = <a_0, a_1>$ can be written in the form $d^* = ha_0 + ka_1$ where h and k are integers. If $a_0 = a_1$, $d^* = a_0 = a_1$, so $h = 1$, $k = 0$ will do. If $a_0 = a_1$, that is, $a_0 > a_1$, then $d^* = a_{n+1}$ where a_{n+1} is obtained by the Euclidean algorithm as above and, using the last two equations of (3.4),

$$\begin{aligned} a_{n+1} = a_{n-1} - q_{n-1}a_n &= a_{n-1} - q_{n-1}(a_{n-2} - q_{n-2}a_{n-1}) \\ &= -q_{n-1}a_{n-2} + (1 + q_{n-1}\,q_{n-2})a_{n-1}\,. \end{aligned}$$

Next we employ the $(n - 2)$nd equation of (3.4), not written down in the text, to replace a_{n-1} and so on until we have used the first equation of (3.4) so as to obtain an expression for $d^* = a_{n+1}$, which is of the required form. For example, for the numerical case considered previously,

$$\begin{aligned} 4 &= 12 - 1 \cdot 8 = 12 - (20 - 1 \cdot 12) = 2 \cdot 12 - 20 \\ &= 2(72 - 3 \cdot 20) - 20 = 2.72 - 7.20 \end{aligned}$$

so that 2 and -7 are suitable values for h and k.

The following theorem expresses a characteristic property of prime numbers. It is important for the sequel but it is also interesting in itself because it characterizes a prime number not in terms of what it is divisible by, as in the original definition, but in terms of its properties as a divisor.

Theorem. Let p be a positive integer greater than 1. In order that p be prime it is necessary and sufficient that whenever p divides a product of positive integers, ab, then p divides at least one of the two numbers a and b.

We write $p|a$ if p divides a, $p \nmid a$ if p does not divide a. Now if p is not prime then $p = qr$ where q and r are positive integers greater than 1 but smaller than p. (Why?) It follows that p divides neither q nor r although $p|qr$ (that is, $p|p$). This shows that the condition of the theorem is sufficient.

Conversely, suppose that p is prime. Then its only divisors are 1 and p, and so these are the only divisors which p can have in common with any other positive integer a. Thus, if $p \nmid a$ then $<p,a> = 1$ and so there exist integers h and k such that

$$ha + kp = 1 . \tag{4.1}$$

Suppose now that p divides the product of a and of another positive integer b, so $ab = cp$ for some positive integer c. Multiplying (4.1) by b we obtain

$$hab + kpb = b$$

or

$$hcp + kbp = b$$
$$(hc + kb)p = b$$

which proves that p divides b. This shows that the condition of the theorem is necessary, and completes its proof.

The following immediate consequence of the theorem just proved will be used later.

Corollary. If a prime number divides a product of several factors then it divides one of the factors. (Why?)

5. PRIME FACTORIZATION

By *the prime factors* of a positive integer a we mean the prime numbers which divide a. As p can divide a only if $p \leq a$, we may find the prime factors of a by checking for all prime numbers up to a whether or not they divide a. Every positive integer a which is greater than 1 can be represented as a product of primes. To see this, we proceed as follows. We begin with the smallest prime number, 2, and divide a by 2. If we obtain a remainder (that is, if a is odd) then we go on to the next prime,

3. If a is divisible by 2 (that is, if a is even) then we write $a = 2a'$ and continue the process with a'. For example, let $a = 120$. Dividing by 2, we obtain 60, that is, $120 = 2 \cdot 60$. Taking 60 and dividing it by 2 we obtain 30, $60 = 2 \cdot 30$, $120 = 2 \cdot 2 \cdot 30$. Similarly, $30 = 2 \cdot 15$, $120 = 2 \cdot 2 \cdot 2 \cdot 15$. Next, we consider 15 and find that it is not divisible by 2. We go on to 3 and find that $15 = 3 \cdot 5$. 5 is prime, so it is not divisible by any other prime. Hence

$$120 = 2 \cdot 2 \cdot 2 \cdot 3 \cdot 5$$

which shows that 120 is a product of primes.

Going back to the general case, we observe that the procedure terminates after a finite number of steps. (Why?) Thus, we obtain a representation of a as a product of powers of primes $p_1, \ldots, p_k$, $k \geq 1$, $p_1 < p_2 < \cdots < p_k$, $n_i \geq 1$, $i = 1, \ldots, k$, so

$$(5.1) \qquad a = p_1^{n_1} p_2^{n_2} \cdots p_k^{n_k}.$$

Although we have obtained $p_1, \ldots, p_k, n_1, \ldots, n_k$ by a definite procedure it is conceivable that there exists some other product of prime numbers which is equal to the same number a. This is indeed possible as we see by changing the order of the primes in the product (provided it contains at least two different primes). However, if we rule out this possibility by stipulating that the primes appear in the product in their natural order, then it can be shown that for given $a > 1$, the product on the right-hand side of (5.1) is unique.

Indeed, suppose that at the same time

$$(5.2) \qquad a = q_1^{m_1} q_2^{m_2} \cdots q^{m_l}$$

where $l \geq 1$, $q_1, q_2, \ldots, q_l$ are prime, $q_1 < q_2 < \cdots < q_l$, $m_i \geq 1$, $i = 1, \ldots, l$. We are going to show that

$$(5.3) \qquad k = l,\ p_i = q_i,\ n_i = m_i,\ i = 1, \ldots, k.$$

For this purpose, we shall use the principle of the smallest number. If there is a number $a (a > 1)$ for which the assertion is not true, then there exists a smallest number of this kind,

$$a = p_1^{n_1} p_2^{n_2} \cdots p_k^{n_k} = q_1^{m_1} q_2^{m_2} \cdots q_k^{m_k}.$$

Since p_1 divides a it must divide one of the q_i (by the corollary given at the end of section 4). But, if $p_1 | q_i$ then $p_1 = q_i$ since q_i has no divisors other than 1 and itself. If $a = p_1 b$, we then have

$$(5.4) \qquad b = p_1^{n_1 - 1} p_2^{n_2} \cdots p_k^{n_k} = q_1^{m_1} q_2^{m_2} \cdots q_i^{m_i - 1} \cdots q_l^{m_l}$$

where $b < a$. Now if $b = 1$ then $k = l = 1$, $p_1 = q_1 = a$ so that a satisfies our assertion, contrary to assumption. If $b > 1$ then b satisfies the assertion to be proved since it is smaller than a.

Suppose now that $n_1 > 1$. Then the number of powers of the p_i in b is k, and these p_i must coincide with the q_i on the right-hand side of (5.4). Hence, in particular, $p_1 = q_1$ and so $i = 1$. Also, from (5.3) as applied to b we then obtain $k = l$; $n_1 - 1 = m_1 - 1$ and so $n_1 = m_1$; and $n_2 = m_2, \ldots, n_k = m_k$, that is, a also satisfies the assertion, again contrary to assumption. Finally, if $n_1 = 1$, then p_1 cannot be a divisor of b and so $m_i - 1 = 0$, $m_i = 1$. In this case again $i = 1$, for if not, then $p_2 = q_1$, by applying the assertion to b, and on the other hand, $q_1 < q_i = p_1$. But this is impossible since $p_1 < p_2 = q_1$, and the two inequalities contradict one another. Hence, in this case $n_1 = m_1 = 1$, $p_1 = q_1$, and, by the representations of b in terms of the p_i and q_i respectively, $k = l$, $p_i = q_i$, $2 \leq i \leq k$. This completes the argument.

We should mention here that it is quite common to gloss over special cases, such as the case $m_1 = n_1 = 1$, both in the text of a proof and in the notation used. In some cases the formal validity of an argument can still be saved by a felicitous wording of the text, but it may well be safer to look a special case firmly into the eye. As far as the notation is concerned, you may have noticed that we have ourselves succumbed to the temptation of taking certain liberties with it. For example, if in (5.4), $i = 1$, then the right-hand side is strictly incorrect since we should then have omitted the term $q_1^{m_1}$.

We sum up the principal result of this section as follows:

Prime Power Factorization Theorem. Every positive integer $a > 1$ can be represented as a product of positive powers of prime numbers, where the primes are taken in their natural order,

$$a = p_1^{n_1} p_2^{n_2} \cdots p_k^{n_k},$$
$$k \geq 1,\ p_1 < p_2 < \cdots < p_k,\ n_i \geq 1,\ 1 \leq i \leq k\,.$$

This representation of a is unique.

6. FACTORIZATION OF INTEGERS

The discussion in section 5 dealt with positive integers only. It will be convenient to modify our results slightly so as to include also negative integers. As usual, $|a|$ is the absolute value of an integer, that is, $|a| = a$ if a is positive or 0, and $|a| = -a$ if a is negative. Then $|ab| = |a|\,|b|$ for all integers a and b. (Why?)

Recall that the ring of integers has been denoted by J.

II. Prime Numbers and Composite Numbers

We know that among the natural numbers the only divisor of 1 is 1 itself. Let a be an integer which is a divisor of 1. That is to say, there exists an integer b such that $ab = 1$. But then $|a|\ |b| = 1$, so that $|a|$ is a natural number which is a divisor of 1. It follows that $|a| = 1$ and hence that $a = 1$ or $a = -1$. These two numbers are said to be the *units* of J. However, only 1 is the (uniquely determined) *unit element* of J as defined in Chapter 1, section 3. The introduction of the term *unit* is regrettable because it is likely to cause confusion with *the unit element*, 1, but we feel compelled to adhere to general usage.

Among the integers, a number p which is not a unit is said to be *prime* if its only divisors are $1, -1, p, -p$. You will see that these are indeed divisors of p for every given p. A natural number is a prime in the sense determined previously if and only if it is a prime in the ring of integers. For if an integer a divides p, then $|a|$ divides p. Hence, if p is prime, $|a| = 1$ or $|a| = p$ and so $a = 1$ or $a = -1$ or $a = p$ or $a = -p$. Conversely, if p has a divisor q, $1 < q < p$, then q will remain a divisor of p in the ring of integers and $q \neq 1, q \neq p, q \neq -1, q \neq -p$. If p is a negative integer then p is prime if and only if $|p| = -p$ is prime. By a *factorization of an integer* a into prime powers we mean a representation

$$a = p_1^{n_1} \cdots p_k^{n_k} \tag{6.1}$$

where the p_i are distinct primes in J, and $n_1, \ldots, n_k$ are positive.

It is easy to give examples which show that this representation of a is not unique even if we disregard the order in which the factors appear in the product. For example $6 = 2 \cdot 3 = (-2)(-3)$ are two distinct representations of 6 as a product of the kind specified in (6.1).

In order to overcome the difficulty which is produced by this ambiguity we introduce the following definition. Two integers a and b are called *associated* if $a = \varepsilon b$, where ε is one of the units of J (that is to say, if ε is either 1 or -1). If this condition is satisfied then we write $a \sim b$. Thus $\sim$ denotes a *relation* (the relation of *association*) which has the following properties:

(i) It is *reflexive*, that is to say $a \sim a$ for all integers a. Indeed $a = \varepsilon a$ for $\varepsilon = 1$.

(ii) It is *symmetrical*, that is to say if $a \sim b$ for a and b in J then $b \sim a$. Indeed $a \sim b$ signifies that either $a = b$ or $a = -b$. In the former case $b = a$, in the latter case $b = -a$, and so in either case $b \sim a$.

(iii) It is *transitive*, that is to say for any integers a, b, and c, if $a \sim b$ and $b \sim c$ then $a \sim c$. (Why?)

It follows directly from the definition that $a \sim b$ if and only if $|a| = |b|$. For every integer a there is just one *other* integer associated with it, that is, $-a$, except when $a = 0$, for in this case a is associated only with itself.

Any relation which has the properties of reflexivity, of symmetry, and of transitivity—(i), (ii), (iii), above—is called a relation of *equivalence.* We have, in fact, already used a relation of equivalence, although we did not mention this point explicitly. This is the relation of equality $a = b$. (Why?)

Any integer a which is neither 0 nor 1 nor -1 (that is, which is neither 0 nor a unit) can be written as a product of primes. We have already shown this to be true if a is positive. If a is negative, then by what has just been said $|a| = -a$ is a product of primes, $a = p_1p_2 \cdots p_n$, $n \geq 1$, where the factors p_i are not necessarily different. Hence,

$$a = -a = -p_1p_2 \cdots p_n = (-p_1)p_2 \cdots p_n ,$$

where the right-hand side is the required representation.

For a not equal to 0, 1, or -1, let

$$a = p_1p_2 \cdots p_n = q_1q_2 \cdots q_m,\ n \geq 1,\ m \geq 1 , \tag{6.2}$$

where the p_i and q_i are all primes. We shall show that $n = m$, and that if we pair off the elements of the sequence $(p_1, p_2, \ldots, p_n)$ with the elements of the sequence $(q_1, q_2, \ldots, q_n)$ in a suitable way, $p_i \leftrightarrow q_{j_i}$, $i = 1, 2, \ldots, n$, then $p_i \sim q_{j_i}$. Here, the term "pairing off" implies that as i varies from 1 to n, j_i takes all integer values from 1 to n, though perhaps not in their natural order.

To prove our assertion, we take absolute values in (6.2). This yields

$$|a| = |p_1|\,|p_2| \cdots |p_n| = |q_1|\,|q_2| \cdots |q_m| . \tag{6.3}$$

Now $|a|$ is a positive integer greater than 1 and the p_i, q_i are all primes. It follows that if we multiply equal p_i in (6.3) and range the resulting powers in rising order of the $|p_i|$ then we obtain a representation of a by powers of positive primes as in the prime power factorization theorem of section 5 above. A similar prime power representation is provided by the $|q_i|$. It then follows from the prime power factorization theorem that any positive prime p which appears among the $|p_i|$ (that is to say, is equal to a $|p_i|$) appears also among the q_i and just as often. Pairing off p_i and q_j which correspond to the same prime $p > 0$, that is, such that $|p_i| = |q_j| = p$ we obtain the required correspondence $p_i \leftrightarrow q_{j_i}$ such that $|p_i| = |q_{j_i}|$, that is, such that $p_i \sim q_{j_i}$.

Our conclusions are summed up as follows:

Factorization Theorem for Integers. Every integer a which is different from 0, 1, -1 is a product of primes. And if two representations of a as products of primes are $a = p_1p_2 \cdots p_n$ and $a = q_1q_2 \cdots q_m$ then $n = m$ and we may pair off the element of the two sequences in such a way that corresponding primes are associated. Thus, if the correspondence between the p_i and the q_i is given by $p_i \leftrightarrow q_{j_i}$ then $p_i \sim q_{j_i}$.

7. UNITS AND PRIMES IN INTEGRAL DOMAINS

Some of the notions which were introduced in the preceding section become more transparent if we consider them in relation to an arbitrary integral domain, D, rather than for J alone.

Recall that according to our definition, the unit element, 1, in an integral domain D is different from its zero (neutral element with respect to addition). An element a of D *divides* an element b of D if there exists an element c of D such that $ac = b$. As before, we write $a|b$ if a divides b, and $a \nmid b$ if a does not divide b. If a divides 1 then a is said to be a *unit* (or, *invertible*). In particular, 1 is itself a unit.

The product of two units is a unit. (Why?)

Two elements a and b of D are said to be *associated*, and we write $a \sim b$ if there exists a unit ε in D such that $a = \varepsilon b$. The relation of association, $\sim$, is an equivalence; that is to say, it is reflexive, symmetrical, and transitive. (Why?) All units of an integral domain are associated with one another. (Why?) If an element a of D is divisible by an element b of D, then a is divisible also by any number b' which is associated with b.

To prove the last assertion, let $a|b$ and $b' \sim b$. Then $bc = a$ for some c in D, and $b' = \varepsilon b$ where ε is a unit. Thus, there exists an ε' in D such that $\varepsilon'\varepsilon = 1$. Multiplying the equation $bc = a$ by ε we obtain $\varepsilon bc = \varepsilon a$, or $b'c = \varepsilon a$. Multiplying this equation by ε', rearranging and taking into account that $\varepsilon'\varepsilon = 1$, we obtain $b'(\varepsilon'c) = a$. This shows that $b'|a$, as asserted.

For any a and b in D, $a \sim b$ if and only if $a|b$ and $b|a$.

Indeed, if $a \sim b$ then $a = \varepsilon b$ where ε is a unit and so $b|a$. At the same time $b \sim a$ and so $b = \varepsilon' a$ where ε' is a unit and so $a|b$. Conversely, if $a|b$ and $b|a$ and $a = 0$ then $b = 0$, so $a \sim b$. If $a|b$ and $b|a$ and $a \neq 0$ then $b \neq 0$. Also, in that case $b = ca$, $a = c'b$, for certain c and c' in D. Then $b = ca = cc'b$ and so $cc' = 1$. It follows that c and c' are units, and so $a \sim b$, as asserted.

Any element a of D is divisible by all units of D, for if ε is a unit, $\varepsilon'\varepsilon = 1$, then $\varepsilon(\varepsilon'a) = a$ showing that $\varepsilon|a$. If a is neither a unit nor 0 and has no divisors other than (i) the units and (ii) the elements associated with a, then a is said to be *prime* or *a prime.*

You may check that if the integral domain D coincides with the rings of integers J, then the definitions of the relation of association, of units, and of primes given in the present section reduce to those given previously. We may still define the G.C.D. of two integers a, b other than 0 as the *natural* number d such that $d|a$ and $d|b$ and if $d'|a$ and $d'|b$ then $d'|d$.

EXERCISES

1. Find all primes up to 200.

2. Find solutions (x, y) for the following equations or prove that no solution exists.

i. $81x + 3y = 5$
ii. $175x - 13y = 1$
iii. $14x + 217y = 35$.

3. Find the smallest natural number n which is divisible by a and b where

i. $a = 375, b = 245$;
ii. $a = 130, b = 310$;
iii. $a = 2^2 \cdot 5^3 \cdot 6, b = 2 \cdot 5^2 \cdot 6^3$.
(n is called *the lowest common multiple* of a and b.)

4. Prove that $2^n + 1$ is divisible by 3, for odd n.

5. Write $10! = 1 \cdot 2 \cdot 3 \cdot \ldots \cdot 10$ as a product of powers of distinct primes.

III
Fields

1. FIELDS

If we divide one integer by another, we may well find that the result is not an integer. Or, to put it in a different way, if we wish to carry out the operation of division by any number (other than 0) without restriction, then we have to pass from the integers to a larger domain. By extending the integers so as to include all fractions a/b for integers a and b such that $b = 0$, we obtain the rational numbers. The set of *rational numbers* will be denoted by Ra. Ra is a ring and even an integral domain, and in addition to the eleven rules stated in Chapter 1 it satisfies the

Twelfth Rule. For any a which is different from 0 there exists an element b such that $ab = 1$. b is called the *multiplicative inverse* or *reciprocal* of a, and we write $b = a^{-1}$.

Any system F which satisfies the eleven rules of Chapter 1 together with the twelfth rule just added and which contains at least two different elements is called a *field*. You will see without difficulty that a ring F which contains at least two different elements and which, in addition, satisfies the twelfth rule is an integral domain. For if $ab = 0$ for elements a and b of F and if $b = 0$, then by the twelfth rule, there exists an element b^{-1} for which $bb^{-1} = 1$. Then

$$a = a \cdot 1 = a(b \cdot b^{-1}) = (ab)b^{-1} = 0 \cdot b^{-1} = 0$$

where we have used the fact that $0 \cdot c = 0$ for all c. This is a simple consequence of the rules for a ring. (Why?)

Since $a = 0$, we have shown that the eleventh rule of Chapter 1 is satisfied. At the same time, 0 must be different from 1 in F. For if $1 = 0$, then for any element a of F,

$$a = a \cdot 1 = a \cdot 0 = 0$$

so that all elements of F are equal. But this contradicts the assumption that F contains at least two different elements. It follows that F is a field. In a field, division by a number other than 0 is always possible. For to divide a number a by a number $b \neq 0$ is to find a number c such that $bc = a$. Now let $c = b^{-1}a$. Then

$$bc = b(b^{-1}a) = (bb^{-1})a = 1 \cdot a = a$$

so that the required number exists. It is customary to write $c = a/b$ as for ordinary fractions. (Why is this all right?)

We shall assume that you are familiar with the system of real numbers and know that it satisfies the twelve rules (and, naturally, contains more than one element); that is, that it constitutes a field. It also helps if you know about the complex numbers, and that they constitute a field, but this is not essential for the understanding of most of this book. We shall denote the field of real numbers by *Re* and the field of complex numbers by C. Then *Ra* is included in *Re*, and *Re* is included in C, which is indicated in symbols by $Ra \subset Re$ and $Re \subset C$. The fields in which we shall be chiefly interested for the remainder of this book are extensions of the field of rational numbers which are contained in the field of complex numbers or, more particularly, the field of real numbers.

2. QUADRATIC FIELDS

Let

$$ax^2 + bx + c = 0 \tag{2.1}$$

be any quadratic equation with rational coefficients, where $a = 0$. We know from high school algebra that the roots (solutions) of this equation are given by

$$\alpha = \frac{1}{2a}\left[-b \pm \sqrt{(b^2 - 4ac)}\right]. \tag{2.2}$$

Thus, the roots of (2.1) are equal if $b^2 - 4ac = 0$, real and different if $b^2 - 4ac > 0$, and complex and different if $b^2 - 4ac < 0$. α is rational if and only if the square root of $b^2 - 4ac$ is rational. The following theorem shows that there exists real numbers which are irrational (not rational).

Theorem. The square root of 2 is irrational.

By "the square root of 2" without further qualification we mean the positive square root of 2. Without explicit reference to irrational numbers, we may restate the assertion of the theorem by saying that the square of any rational number a is different from 2, or $(m/n)^2 \neq 2$, that is

$$m^2 \neq 2n^2$$

where m and n are any integers, $n \neq 0$.

Since $m^2 = (-m)^2$ for any integer m, it is clearly sufficient to prove our assertion for all natural numbers m and n. Suppose that there exist natural numbers m and n, $n > 0$ (and hence $m > 0$) such that

$$m^2 = 2n^2 .$$

By the prime power factorization theorem there exists a natural number k such that m is divisible by 2^k but not by any higher power of 2. It follows that m^2 is divisible by 2^{2k} but not by any higher power of 2. (Why?) Similarly, there exists a natural number l such that n is divisible by 2 but not by any higher power of 2, and $2n^2$ is divisible by 2^{2l+1} but not by any higher power of 2. But since $m^2 = 2n^2$, it follows that $2k = 2l + 1$, so $2k$ is a natural number which is both odd and even. This is impossible. (Why?) Accordingly our assertion is proved.

The square roots of 2 are the roots of the quadratic equation $x^2 - 2 = 0$, so we have shown that the roots of a quadratic equation may be irrational.

Going back to the general case, multiplication by a suitable natural number shows that any number which is a root of a quadratic equation with rational coefficients a, b, c is also a root of a quadratic equation with integer coefficients. For example, the roots of

$$\tfrac{5}{3}x^2 - \tfrac{7}{4}x + 1 = 0$$

are the same as the roots of

$$20x^2 - 21x + 12 = 0$$

which is obtained by multiplying the former equation by $3 \cdot 4 = 12$. Accordingly, we may suppose that a, b, and c in (2.1) are integers.

Let α be a root of a specified equation (2.1), given by (2.2), such that α is irrational. α is thus a (real or complex) number outside Ra. We consider the set of all numbers

$$p + q\alpha \tag{2.3}$$

where p and q are rational (that is, elements of Ra). We denote this set by $Ra(\alpha)$.

Notice that different expressions (2.3) denote different numbers. That is to say

$$p_1 + q_1\alpha = p_2 + q_2\alpha\,, \tag{2.4}$$

where p_1, q_1, p_2, q_2 are rational, only if $p_1 = p_2$ and $q_1 = q_2$. For if $q_1 \neq q_2$ then, by (2.4)

$$\alpha = \frac{p_1 - p_2}{q_1 - q_2}$$

so that α would be rational, contrary to assumption. Hence, $q_1 = q_2$ and so $q_1\alpha = q_2\alpha$. Subtracting this equation from (2.4), we obtain $p_1 = p_2$ proving our assertion.

The sum of two elements of $Ra(\alpha)$ *is an element of* $Ra(\alpha)$.

For if $r_1 = p_1 + q_1\alpha$ and $r_2 = p_2 + q_2\alpha$ as in (2.3) then $r_1 + r_2 = (p_1 + q_1\alpha) + (p_2 + q_2\alpha) = (p_1 + p_2) + (q_1 + q_2)\alpha$. You will see that the right-hand side is of the form (2.3) so that $r_1 + r_2$ belongs to $Ra(\alpha)$. This shows that $Ra(\alpha)$ satisfies the fifth rule. Similarly,

The difference of two elements of $Ra(\alpha)$ *is an element of* $Ra(\alpha)$.

In particular, if r belongs to $Ra(\alpha)$ then $-r$ belongs to $Ra(\alpha)$. Thus $Ra(\alpha)$ satisfies our fourth rule.

The product of two elements of $Ra(\alpha)$ *is an element of* $Ra(\alpha)$.

For if $r_1 = p_1 + q_1\alpha$ and $r_2 = p_2 + q_2\alpha$ as in (2.3), then

$$r_1r_2 = (p_1 + q_1\alpha)(p_2 + q_2\alpha) = p_1p_2 + p_1q_2\alpha + q_1p_2\alpha + q_1q_2\alpha^2\,. \tag{2.5}$$

But α is a root of (2.1) so

$$a\alpha^2 + b\alpha + c = 0$$

or

$$\alpha^2 = -\frac{b}{a}\alpha - \frac{c}{a}\,. \tag{2.6}$$

Substituting this expression for α^2 on the right-hand side of (2.5) and rearranging, we obtain

$$r_1r_2 = \left(p_1p_2 - \frac{q_1q_2c}{a}\right) + \left(p_1q_2 + q_1p_2 - \frac{q_1q_2b}{a}\right)\alpha\,,$$

which is of the form (2.3). Hence, r_1r_2 belongs to $Ra(\alpha)$, $Ra(\alpha)$ satisfies the sixth rule.

Both 0 and 1 belong to $Ra(\alpha)$ for 0 can be written as

$$0 = 0 + 0 \cdot \alpha\,.$$

and 1 as

$$1 = 1 + 0 \cdot \alpha\,.$$

Also $r + 0 = r$ and $r \cdot 1 = r$ in $Ra(\alpha)$ since this is true in the wider field (the field of real numbers or the field of complex numbers) in which r, 0, and 1 are considered. Similarly, all the other rules from the first to the eleventh are satisfied in $Ra(\alpha)$ because they are satisfied in Re or C. We conclude that $Ra(\alpha)$ is an integral domain.

But $Ra(\alpha)$ is even a field, that is to say it satisfies the twelfth rule also. For let $r = p + q\alpha$ as in (2.3) and suppose that $r \neq 0$, that is, that p and q are not both 0. We have to find a number $p' + q'\alpha$, where p' and q' are rational such that $(p + q\alpha)(p' + q'\alpha) = 1$, that is, such that

$$pp' + pq'\alpha + qp'\alpha + qq'\alpha^2 = 1 .$$

Now by (2.6) this will be the case if

$$\left(pp' - \frac{qq'c}{a}\right) + \left(pq' + qp' - \frac{qq'b}{a}\right)\alpha = 1$$

and this again is true if

$$\begin{aligned} pp' - \frac{qq'c}{a} &= 1 \\ pq' + qp' - \frac{qq'b}{a} &= 0 . \end{aligned} \tag{2.7}$$

We may regard (2.7) as a system of linear equations for the unknowns p' and q', and our problem is to show that the system has a solution. Rearranging, we get

$$\begin{aligned} app' - cqq' &= a \\ aqp' + (ap - bq)q' &= 0 . \end{aligned} \tag{2.8}$$

Multiplying the first of these equations by q and the second by p and subtracting the first from the second equation, we obtain

$$(ap^2 - bpq + cq^2)q' = aq .$$

Assuming that $ap^2 - bpq + cq^2$ is different from 0, we may divide by it and obtain

$$q' = -\frac{aq}{ap^2 - bpq + cq^2} \tag{2.9}$$

which is then a rational number since a, b, c, p, and q are rational numbers. Now p and q are not both 0. If $p \neq 0$ then we substitute the expression just obtained for q' in the first equation of (2.8) and determine

p' from it. If $p = 0$, so that $q \neq 0$, then we substitute the expression for q' in the second equation of (2.8) and obtain p' from that equation. In either case, the result is

$$p' = -\frac{ap - bq}{ap^2 - bpq + cq^2} \tag{2.10}$$

which is again rational.

Strictly speaking we have to verify that p' and q' actually satisfy the equation (2.8); this can be done directly, given (2.7) and (2.10). However, it is rather unlikely that we might have found (2.9) and (2.10) by straight guessing, without solving (2.8) by some method or other.

Our proof that the reciprocal of r is contained in $Ra(\alpha)$ is still incomplete because we have not yet shown that the denominator of (2.9) and (2.10) is different from 0. Let us first try to carry out the procedure for a particular case. Suppose $\alpha = \sqrt{2}$, so that the equation (2.1) may be taken as

$$x^2 - 2 = 0$$

that is, with $a = 1, b = 0, c = -2$. Suppose that r is given as $r = 3 - 5\sqrt{2}$. We try to find p' and q' which are rational numbers such that

$$(3 - 5\sqrt{2})(p' + q'\sqrt{2}) = 1$$

that is, such that

$$3p' + 3\sqrt{2}q' - 5\sqrt{2}p' - 10q' = 1$$

or

$$3p' - 10q' + (-5p' + 3q')\sqrt{2} = 1$$

and this is true if

$$3p' - 10q' = 1, -5p' + 3q' = 0\,.$$

Solving these equations for p' and q' we obtain $p' = -\frac{3}{41}$ and $q' = -\frac{5}{41}$, so that no difficulty arises concerning the denominator.

Still taking $\alpha = \sqrt{2}$ let r be general, $r = p + q\sqrt{2}$. Then the condition for p' and q' is

$$(p + q\sqrt{2})(p' + q'\sqrt{2}) = 1$$

or

$$pp' + 2qq' + (pq' + qp')\sqrt{2} = 1$$

which will be satisfied if

$$pp' + 2qq' = 1$$
$$qp' + pq' = 0\,.$$

Solving this system of equations for p' and q' we obtain *formally* (that is, without first checking whether the result is invalidated by a vanishing denominator)

$$p' = \frac{p}{p^2 - 2q^2} \qquad q' = -\frac{q}{p^2 - 2q^2}\,.$$

But the denominator in these expressions is indeed different from 0. For if $p^2 - 2q^2 = 0$ then $2 = (p/q)^2$ so that 2 would be the square of a rational number and we have shown that this is not the case. Thus, we have relied on the irrationality of $\sqrt{2}$ in order to settle our problem. This gives us a clue of what to do in the general case.

In that case, the denominator of (2.9) and (2.10) is $ap^2 - bpq + cq^2$. Suppose now that

$$ap^2 - bpq + cq^2 = 0\,. \tag{2.11}$$

If at the same time $q = 0$, this yields $ap^2 = 0$, hence $p = 0$, contrary to assumption. So we may divide by q^2 and obtain

$$a\left(-\frac{p}{q}\right)^2 + b\left(-\frac{p}{q}\right) + c = 0\,.$$

This shows that $-p/q$ is a root of equation (2.1), that is, that equation possesses a rational root. On the other hand, α is an irrational root of (2.1) by assumption. However, the two roots of (2.1) are given by (2.2). You may deduce from that equation, or in other ways, that a quadratic equation which has a rational root cannot have an irrational root. (Why?) This shows that (2.11) is impossible and completes the proof that $Ra(\alpha)$ is a field. Such a field is called *quadratic*.

We shall reserve the name "quadratic equation" for equations (2.1) in which $a \neq 0$.

Theorem. If α is a real or complex number which is a root of a quadratic equation with rational coefficients then every element of $Ra(\alpha)$ is a root of a quadratic equation with rational coefficients.

Proof. Let β be a number which belongs to $Ra(\alpha)$. Then $\beta = p + q\alpha$ where p and q are rational. If $q = 0$ then β is actually rational $(=p)$ so it is certainly the root of linear equations, such as $x - p = 0$, and also of quadratic equations, such as $x^2 - 2xp + p^2 = 0$. Suppose now that $q \neq 0$ and write $\alpha = (1/q)\beta - p/q$. Substitution of this expression for x in (2.1) yields

$$a\left(\frac{1}{q}\beta - \frac{p}{q}\right)^2 + b\left(\frac{1}{q}\beta - \frac{p}{q}\right) + c = 0\,.$$

Expanding and rearranging we obtain

$$\frac{a}{q^2}\beta^2 + \left(-\frac{2ap}{q^2} + \frac{b}{q}\right)\beta + \left(c - \frac{bp}{q}\right) = 0\,.$$

This shows that β is a root of a quadratic equation with rational coefficients and hence also a root of a quadratic equation with integer coefficients.

For example, $\beta = 3 - \sqrt{2}$ is an element of the field $Ra(\sqrt{2})$. Since $\sqrt{2}$ is a root of the equation $x^2 - 2 = 0$ and $\sqrt{2} = 3 - \beta$, we have $(3 - \beta)^2 - 2 = 0$, that is, β is a root of the quadratic equation $x^2 - 6x + 7 = 0$.

Let α be irrational and a root of a quadratic equation as before, and let $\beta = p + q\alpha$ where p and q are rational and $q \neq 0$. Then we claim that $Ra(\beta) = Ra(\alpha)$.

You will see without difficulty that every element γ of $Ra(\beta)$ belongs also to $Ra(\alpha)$. For if γ belongs to $Ra(\beta)$ then there exist rational numbers p' and q' such that $\gamma = p' + q'\beta$. But then

$$\gamma = p' + q'(p + q\alpha) = (p' + q'p) + q'q\alpha = p'' + q''\alpha$$

where $p'' = p' + q'p$ and $q'' = q'q$ are both rational. This shows that γ belongs to $Ra(\alpha)$.

Conversely, since $\beta = p + q\alpha$, where $q \neq 0$, it follows that $\alpha = (1/q)\beta - p/q$. Therefore, if some γ belongs to $Ra(\alpha)$ and, accordingly, can be written as $\gamma = p' + q'\alpha$, then $\gamma = p' + q'\ [(1/q)\beta - p/q] = [(p' - q'p/q) + (q'/q)\beta]$ so that γ is in $Ra(\beta)$. Evidently β is irrational in the case under consideration. (Why?)

Suppose that α is a root of (2.1) so that $\alpha = (-b/2a) \pm (1/2a)\sqrt{(b^2 - 4ac)}$. You may check that the following argument, which is carried out for the case that the $+$ sign in $\pm$ applies, is equally valid if the $-$ sign is chosen. Putting $\beta = \sqrt{(b^2 - 4ac)}$, we then find that $\beta = b + 2a\alpha$, and so $Ra(\beta) = Ra(\alpha)$. In general, β is simpler than α because β is a square root of an integer (a, b, and c, being taken as integers, as before). And since we do not get anything more general by taking α as in (2.2), we shall suppose from now on that the α in $Ra(\alpha)$ is itself the square root of an integer. If that integer is positive or zero then β is real; if not then β is complex.

3. CONGRUENCE

We refer to the system represented by Figures 2, 3 in Chapter 1. This system was obtained by adding and multiplying the numbers 0, 1, 2, 3, 4 and taking remainders with respect to 5. Clearly, it contains more than two different elements. However, a direct check shows that the twelfth rule is also satisfied, so that the system under consideration constitutes a field. In order to deal with such systems efficiently, we introduce the notion of *congruence.*

Considering the ring of integers, J, let n be a positive integer greater than 1 and let a and b be arbitrary integers (which may coincide). We say that *a is congruent to b modulo n;* in symbols: $a \equiv b(n)$, or $a \equiv b$ mod. n if the difference $a - b$ is divisible by n, which is to say if $(a - b)/n$ is an integer.

The relation of congruence modulo a given $n > 1$ is an equivalence. Indeed

(i) it is reflexive, $a \equiv a(n)$, since $(a - a)/n$ is an integer;
(ii) it is symmetrical, $a \equiv b(n)$ entails $b \equiv a(n)$, for if $(a - b)/n$ is an integer then $(b - a)/n = -(a - b)/n$ also is an integer; and
(iii) it is transitive, $a \equiv b(n)$ and $b \equiv c(n)$ together entail $a \equiv c(n)$, for if $(a - b)/n$ and $(b - c)/n$ are integers then $(a - c)/n = (a - b)/n + (b - c)/n$ also is an integer.

An integer a is either even or odd according as $a \equiv 0(2)$ or $a \equiv 1(2)$.

For given n, we divide J into classes $A, B, C, \ldots$ in the following way, so that no two classes have an element in common (and, as implied, such that all the classes together exhaust J). The numbers a and a' of J shall belong to the same class A if $a \equiv a'(n)$ while if $a \equiv a'(n)$ is not true then a and a' shall belong to different classes. (If $a \equiv a'(n)$ is not true we shall write also $a \not\equiv a'(n)$). You will have no difficulty in verifying that such a partition into classes is indeed possible. If k is any number between 0 and $k - 1$ (both of these included), then the class which includes k and which will be denoted by A_k consists of all numbers a which can be written in the form $a = k + jn$ where j is an arbitrary integer. The classes $A_0, A_1, \ldots, A_{n-1}$ together exhaust J.

We now regard the classes $A_0, \ldots, A_{n-1}$ as the elements of a new mathematical structure R_n, in which the operations of addition and multiplication are defined as follows.

For any A_j, A_k in R_n, let a be any element of A_j and let b be any element of A_k. Then we define the sum $A_j + A_k$ as the class A_m which contains $a + b$. In order to be sure that this is a good definition, we have to verify that, for given A_j and A_k, the sum A_m is independent of the particular choice of a and b. Thus, if a and a' belong to A_j, and b and b' belong to A_k, then we have to show that $a + b$ belongs to the same equivalence class as $a' + b'$. The assumption is that $a \equiv a'(n)$ and $b \equiv b'(n)$, that is, that both $a - a'$ and $b - b'$ are divisible by n. But if so, then $a + b - (a' + b') = (a - a') + (b - b')$ is divisible by n. This shows that $a + b$ and $a' + b'$ belong to the same equivalence class A_m.

To find the subscript m for given j and k, we remember that the numbers j and k belong to the equivalence classes A_j and A_k respectively. Thus A_m should contain the sum $j + k$, and also the (unique) number between 0 and $n - 1$ (these numbers included) which differs from $j + k$ by a multiple of n. But this is precisely the remainder of $j + k$ after division by n. This number must be m. For example, if $n = 5$, we have the classes A_0, A_1, A_2, A_3, A_4. Examples for the addition of these classes are

$$A_0 + A_2 = A_2, \qquad A_1 + A_3 = A_4, \qquad A_2 + A_3 = A_0, \qquad A_3 + A_4 = A_2.$$

You may check that the subscripts on the right-hand side are given precisely by Figure 2 of Chapter 1. In fact, the procedure described above for finding the class which is to be the sum of two given classes of R_n shows that quite generally

$$A_j + A_k = A_{j \oplus k}$$

where the operation $\oplus$ is defined as in Chapter 1.

In order to define the product of two classes A_j and A_k, we again choose two arbitrary elements a and b in A_j and A_k respectively, and we take A_jA_k to be the class A_m which contains the number ab. This definition is unique, for if a' and b' also belong to A_j and A_k respectively, then $a - a'$ and $b - b'$ are divisible by n,

$$a - a' = \lambda n \qquad b - b' = \mu n$$

where λ and μ are integers. Hence

$$ab - a'b' = (a - a')b + a'(b - b') = (\lambda b + \mu a')n$$

so that the class A_m which contains ab contains also $a'b'$, the outcome is indeed independent of the particular choice for a and b.

If we take j for a and k for b (since these numbers belong to A_j

and A_k respectively) then we find that A_m contains all numbers of the form $jk - \lambda n$. Now let the result of dividing jk with remainder by n be

$$jk = \lambda n + \mu, \text{ when } 0 \leq \mu \leq n - 1.$$

Then $\mu = jk - \lambda n$ belongs to A_m. However, only one number between 0 and $n - 1$ (these numbers included) can belong to a class. For the difference between two distinct numbers between 0 and $n - 1$ cannot be divisible by n, and so the two numbers cannot be congruent. But m belongs to A_m and μ belongs to A_m, and so we may conclude that $\mu = m$. This shows that

$$A_j A_k = A_{j \odot k}$$

where the operation $\odot$ is defined as in Chapter 1.

You should have no difficulty in verifying that with these definitions R_n is a ring with unit element, that is, A_1. The zero of R_n is A_0.

4. ISOMORPHISM

Recall that a one-to-one mapping (bijection) from a set S' to a set S'' of any objects whatever is a function $y = \phi(x)$ whose argument values range over all of S', and whose functional values belong to S'' in such a way that every element b of S'' occurs at least once as function value, $b = \phi(a)$ for some a in S'; and every element b of S'' occurs as a function value not more than once. That is to say if $b = \phi(a_1)$ and $b = \phi(a_2)$ where a_1 and a_2 are in S' by assumption, then $a_1 = a_2$. This relation between S' and S'' is indicated also by $S' \overset{\phi}{\leftrightarrow} S''$; and if $b = \phi(a)$ we may write $a \overset{\phi}{\leftrightarrow} b$, (or $a \leftrightarrow b$, omitting the symbol ϕ if the function is taken for granted).

A ring R' is said to be *isomorphic to* a ring R'' if there is a one-to-one mapping 0 from R' to R'' such that the following two conditions are satisfied:

(i) If $b_1 = \phi(a_1)$ and $b_2 = \phi(a_2)$ then $b_1 + b_2 = \phi(a_1 + a_2)$
(ii) If $b_1 = \phi(a_1)$ and $b_2 = \phi(a_2)$ then $b_1 \cdot b_2 = \phi(a_1 \cdot a_2)$

A function or mapping ϕ which enjoys these properties is said to be *an isomorphism*.

If R' is isomorphic to R'' then R'' is isomorphic to R'. (Why?)

Some remarks on notation may be useful here. When discussing two rings, such as R' and R'', it is customary to denote the zeros of both rings by the same symbol, 0, and frequently it turns out to be irrelevant whether or not we suppose that the zeros of the two rings actually coincide. Similarly, if both R' and R'' have unit elements (neutral elements with respect to multiplication) it is customary to denote both by 1. On the other hand, we may denote the operations of addition and of multiplication in both rings by the same symbols as above, $+$ and $\cdot$, (usually omitting the dot, in accordance with custom); or we may use different symbols for the operations in the two rings, $+$ or $\cdot$ for the appropriate operations in one ring, $\oplus$ and $\odot$ in the other. In that case, the conditions in (i) and (ii) have to be replaced by $b_1 \oplus b_2 = \phi(a_1 + a_2)$ and $b_1 \odot b_2 = \phi(a_1 \cdot a_2)$, respectively.

If R' is isomorphic to R'' then the zeros of the two rings correspond, $0 = \phi(0)$.

To see this, let $b = \phi(0)$. Then we have to show that b is the neutral element with respect to addition in R''. That is to say, for any d in R'', $d + b = d$. Now, by assumption, there exists an element c in R' such that $d = \phi(c)$. Then by (i) above, $d + b = \phi(c + 0) = \phi(c)$, that is, $d + b = d$. This proves our assertion.

If R' contains a unit element, 1, *then R'' contains a unit element,* 1, *and* $1 = \phi(1)$.

Suppose that $b = \phi(1)$. We have to show that if d is any element of R'' then $d \cdot b = d$. Now if $d = \phi(c)$ then $d \cdot b = \phi(c)\phi(1) = \phi(c \cdot 1) = \phi(c)$, where we have made use of (ii). But $\phi(c) = d$ and so $d \cdot b = d$, as required.

If $b = \phi(a)$ then $-b = \phi(-a)$.

For if $d = \phi(-a)$ then $b + d = \phi(a) + \phi(-a) = \phi(a - a) = \phi(0) = 0$. Thus, $b + d = 0$, d is the inverse of b with respect to addition, $d = -b$, as asserted.

Our results show that as far as the operations of addition and multiplication are concerned, isomorphic rings are indistinguishable.

For given $n > 1$, the ring R_n is isomorphic to the ring which consists of the numbers $0, 1, \ldots, n - 1$, in which addition and multiplication are given by the operations $\oplus$ and $\odot$ defined in Chapter 1. The required mapping ϕ is provided by $k = \phi(A_k)$, $k = 0, 1, \ldots, n - 1$. (Why?) Accordingly, we may, if we like, consider the ring R_n rather than the set $\{0, 1, \ldots, n - 1\}$ with the operations $\oplus$ and $\odot$.

Theorem. R_n is a field if and only if n is prime.

Proof. Suppose that n is not prime. Then there exist numbers j and k, both greater than 1 but smaller than n such that $jk = n$. Consider the elements A_j and A_k of R_n. They are both different from the zero of R_n which is A_0. However, $A_jA_k = A_{j \odot k} = A_0$, so that R_n contains two elements which are both different from zero, but their product is zero. This shows that the twelfth rule is not satisfied. In other words, if R_n is a field, then n must be prime.

Conversely, suppose that n is prime. Suppose at the same time that there exist elements of R_n, A_j and A_k, such that neither A_j nor A_k is equal to zero, yet A_jA_k is equal to zero; that is to say, $A_jA_k = A_0$. But this means that jk is divisible by n. According to the theorem of Chapter II, section 4, this in turn implies that either j or k is divisible by n, which is impossible since both j and k are positive integers smaller than n.

Thus, for prime n, the R_n provide examples of fields which contain a finite number of elements only. A significant difference exists between these fields on one hand, and the fields contained in the field of complex numbers—for example, the fields Ra and Re and the fields $Ra(\alpha)$—introduced in the preceding sections. For the latter, the sums $1 + 1$, $1 + 1 + 1$, $1 + \ldots + 1$ (1 added to itself a finite number of times) are all different from zero, they are, in fact, all positive integers. On the other hand, if in R_n, n a prime number, we add 1 a finite number of times to itself then we may obtain zero. More particularly, in R_n,

$$1 + 1 + \ldots + 1 \ (n \text{ times})$$

is equal to A_0, that is, it is equal to the zero of the field.

Definition. For any field F the smallest possible integer, p, which is such that $1 + 1 + \ldots + 1$ (p times) is equal to zero, is called the *characteristic* of the field. If there is no positive integer q such that $1 + 1 + \ldots + 1$ (q times) is equal to 0, then we say that F is of characteristic 0. The rational numbers, the complex numbers, and all fields contained in the complex numbers are of characteristic 0. The field R_n introduced above is of characteristic n. If a field is not of characteristic 0, then its characteristic, p, must be a prime number. (Why?)

The notion of congruence can be given the following concrete representation. Suppose that points have been marked off on a tape at unit distance from one another. We choose one of the marks as 0 and we may then assign to the other marks the numbers 1, 2, 3, . . . on moving from 0 to the right, and the numbers -1, -2, -3, . . . on moving from 0 to the left.

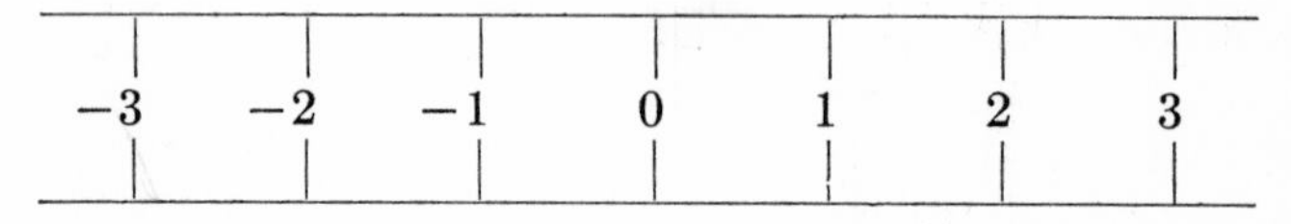

Fig. 5.

We imagine, ideally, that the tape extends indefinitely in both directions. Given an integer $n > 1$, take a circular cylinder of radius $r = n/2\pi$, and wrap the tape around it in the natural way (that is, in a direction normal to the generators of the cylinder). Since the circumference of the cylinder is $2\pi r = n$, the distance between marks on the tape which coincide on the cylinder must be a multiple of n; hence the corresponding numbers are congruent modulo n. Conversely, if two numbers are congruent modulo n then the corresponding numbers on the tape coincide on the cylinder.

EXERCISES

1. Prove that $\sqrt{3}$ and $\sqrt[3]{3}$ are irrational.

2. Let S be the set of ordered pairs of real numbers. Define addition of elements of S "by components"

$$(a, b) + (c, d) = (a + c, b + d) .$$

Define multiplication by

i. $(a, b) \cdot (c, d) = (ac, bd)$
ii. $(a, b) \cdot (c, d) = (ac - bd, ad + bc)$.

By just one of these definitions S becomes a field. Which is it?

3. Show that one of the rings introduced in the preceding question is isomorphic to the field of complex numbers.

4. Arrange the numbers 2, 3, 4, 5, 6, 7, 8, 9 in pairs (a, b) so that for each pair $ab \equiv 1(11)$.

5. Show that there is no number x such that

$$3x \equiv 1(6) .$$

For which natural numbers a does there exist an x such that

$$ax \equiv 1(6) ?$$

IV
Integers in Quadratic Number Fields

1. ALGEBRAIC INTEGERS

At the end of Chapter III, section 3, we pointed out that any quadratic field can be obtained as $Ra(\alpha)$ where α is one of the square roots of an integer d, such that α is irrational (that is, such that d has no rational root). We now ask ourselves whether $Ra(\alpha) = Ra(\sqrt{d})$ contains a set, $J(\alpha)$, say, which is related to $Ra(\alpha)$ "just as" the integers are related to the rational numbers. However, we first have to clarify what exactly we mean by "just as" in this case. Let us list some important properties of the system of integers, J.

(i) J is an integral domain which is contained in Ra.
(ii) Every element of Ra is equal to the quotient of two elements of J.
(iii) Every element of J is the root of an equation $x + b = 0$, where b is an integer (that is, in J). Indeed, suppose β is the given element of J. Then $b = -\beta$ is an integer, and $\beta + b = 0$.

This may appear to you as a very complicated way to state a very simple fact. Nevertheless, it turns out to be the clue to a suitable generalization for the notion of an integer.

In the present work we shall consider only numbers which are roots of quadratic equations with rational coefficients (or, equivalently, with integer coefficients). This includes the rational numbers. (Why?)

Definition. The (real or complex) number β is said to be an *algebraic integer* if there exist elements of J, p and q, that is, *integers* in the sense in which this word has been used until now, such that β is a root of the equation $x^2 + px + q = 0$.

By this definition $\sqrt{2}$ is an algebraic integer since it is a root of the equation $x^2 - 2 = 0$; and $3 + \sqrt{2}$ is an algebraic integer since it is a root of the equation $x^2 - 6x + 7 = 0$. But even $\frac{1}{2}(1 - \sqrt{5})$ is an algebraic integer since it is a root of the equation $x^2 - x - 1 = 0$.

Suppose now that β is a rational number. If our notion of an algebraic integer is a genuine generalization of the notion of an integer, we should be able to prove that β is an integer in the ordinary sense if and only if it is an algebraic integer.

Suppose first that β is an integer in the ordinary sense. Then if $p = -\beta$, β is a root of the equation $x + p = 0$ and hence also of the equation $x^2 + px = 0$. Thus β is an algebraic integer. Conversely, suppose that β is a rational number, $\beta = a/b$, where a and b are integers and suppose also that β is a root of the equation $x^2 + px + q = 0$ where p and q are integers. Then we have to show that β is an integer, that is, an element of J.

We observe that one can always replace a fraction a/b by an equal *reduced* fraction a'/b'. That is to say, $a/b = a'/b'$ and, moreover, $(a', b') = 1$, the greatest common divisor of a' and b' is 1. For example $-18/42$ can be replaced by $3/-7$ or by $-3/7$ both of which are reduced. Moreover, we may choose the reduced fraction in such a way that the denominator is positive.

Suppose then that $\beta = a/b$, where a/b is a reduced fraction whose denominator is positive, $(a, b) = 1$ and $b \geq 1$. By assumption, β satisfies the quadratic equation $x^2 + px + q = 0$ and so

$$\left(\frac{a}{b}\right)^2 + p\left(\frac{a}{b}\right) + q = 0 .$$

Multiplying this equation by b^2 we obtain

$$a^2 + pab + qb^2 = 0$$

and so

$$a^2 = b(-pa - b) . \qquad (1.1)$$

Now suppose that b is greater than 1. Then b must possess at least one divisor which is prime, m say. m divides the right-hand side of (1.1) and so it must divide also its left-hand side, a^2. But if m is a prime factor of a^2, it must divide also a. Thus, m divides both a and b, which is contrary to the assumption that the greatest common divisor of a and b is 1.

We may conclude that $b = 1$, and therefore $\beta = a/1 = a$ is an integer.

We have shown that a rational number is an integer in the ordinary sense if and only if it is an algebraic integer. This justifies the dropping of the adjective *algebraic* before *integer*. From now on the term *integer* will refer to arbitrary algebraic integers while the elements of J will be called *rational* integers. (The term *natural number* will retain its meaning.)

Consider any quadratic field $Ra(\alpha) = Ra(\sqrt{d})$, $\alpha = \sqrt{d}$ where d is an integer and α is irrational. The set of integers (that is, algebraic integers, see above) in $Ra(\alpha)$ will be denoted by $J(\alpha)$.

It will be convenient to assume that d is not divisible by the square of any natural number other than 1. In that case, d is said to be *square-free*. The numbers 2, 6, -10, are square-free, 20 is not square-free. d is square-free if and only if it is not divisible by the square of any *prime* number. (Why?)

Any rational integer $a \neq 0$ can be written in the form $a = b^2a'$ where b is a rational number and a' is a square-free rational integer. This is obvious if a is $+1$ or -1 for in that case $b = 1$, $a' = a$ satisfies our conditions. If a is not square-free and $|a| > 1$, then we may express $|a|$ as a product of powers of primes, so

$$|a| = p_1^{n_1}p_2^{n_2} \cdots p_k^{n_k},\ k \geq 1, \text{ and } n_j \geq 1 \text{ for } 1 \leq j \leq k\,.$$

We now write all even n_j in the form $n_j = 2m_j$, and we write all odd n_j as $n_j = 2m_j + 1$ (where m_j is positive if n_j is even and may be 0 if n_j is odd). Put

$$b = p_1^{m_1}p_2^{m_2} \cdots p_k^{m_k}$$

so that $b^2 = p_1^{2m_1}p_2^{2m_2} \cdots p_k^{2m_k}$ divides a (recall that $p_j^0 = 1$). Then $a' = a/b^2$ is square-free since it is divisible by no power of a prime higher than the first. More precisely, $|a'|$ is the product of those primes p_j for which n_j is odd. Thus we have shown that every rational integer n is equal to a product b^2a' where b is a natural number and a' is a square-free rational integer.

For the given d, write $d = b^2d'$ where d' is a square-free rational integer as above, then $\sqrt{d} = \sqrt{(b^2d')} = b\sqrt{d'}$. This shows that $Ra(\sqrt{d'})$ contains $\sqrt{d}$, and hence, contains all other numbers of $Ra(\sqrt{d})$. (Why?) But $\sqrt{d'} = (1/b)\sqrt{d}$, and so $Ra(\sqrt{d})$ contains $\sqrt{d'}$ and all other elements of $Ra(\sqrt{d'})$. This shows that $Ra(\sqrt{d})$ and $Ra(\sqrt{d'})$ coincide. Accordingly, we do not lose anything (that is, we do not miss any quadratic fields) by supposing that the number under the sign of the square root is square-free and we shall do so from now on.

Thus, consider $Ra(\alpha) = Ra(\sqrt{d})$ where d is square-free. Any element β of $Ra(\sqrt{d})$ can be written as $\beta = a + b\sqrt{d}$ where a and b are rational numbers. Consider the following product, where x is a variable

$$\begin{aligned} P(x) &= [x - (a + b\sqrt{d})]\,[x - (a - b\sqrt{d})] \\ &= (x - a + b\sqrt{d})(x - a - b\sqrt{d}) \\ &= (x - a)^2 - (b\sqrt{d})^2 = x^2 - 2ax + (a^2 - b^2d) = 0\,. \end{aligned}$$

Evidently, $P(\beta) = [\,\beta - (a + b\sqrt{d})][\beta - (a - b\sqrt{d})] = 0[\beta - (a - b\sqrt{d})\,] = 0$ so that β is a root of the equation

$$x^2 - 2ax + (a^2 - b^2d) = 0\,. \tag{1.2}$$

We shall show that if β is an integer (that is to say, an algebraic integer) then the numbers $2a$ and $a^2 - b^2d$, which appear in (1.2) must be integers (that is to say, rational integers since they are rational numbers).

Suppose that $\beta = a + b\sqrt{d}$ is an integer. If β is rational then $b = 0$, $\beta = a$, so that in this case a is a rational integer, and $2a$ and $a^2 - b^2d = a^2$ are both rational integers. Suppose now that β is not rational or, which is the same, that $b \neq 0$. By assumption, β satisfies an equation

$$x^2 + px + q = 0 \tag{1.3}$$

where p and q are rational integers. We claim that $p = -2a$, $q = a^2 - b^2d$. In other words, the polynomials on the left-hand side of (1.2) and (1.3) coincide. Substituting β in (1.2) and (1.3) we obtain

$$\begin{aligned} &\beta^2 - 2a\beta + (a^2 - b^2d) = 0 \\ &\beta^2 + p\beta + q = 0\,. \end{aligned}$$

Subtracting the first from the second equation we have

$$(p + 2a)\beta + (q - a^2 + b^2d) = 0\,.$$

Substituting $a + b\sqrt{d}$ for β in this equation,

$$(p + 2a)(a + b\sqrt{d}) + (q - a^2 + b^2d) = 0\,,$$

or, rearranging,

$$a(p + 2a) + (q - a^2 + b^2d) + b(p + 2a)\sqrt{d} = 0\,. \tag{1.4}$$

Now, by Chapter III, section 2, an equation $p_1 + q_1\sqrt{d} = 0 = 0 + 0\sqrt{d}$ with irrational $\sqrt{d}$ and rational p_1 and q_1 is possible only if $p_1 = q_1 = 0$. This can also be shown more directly by noting that if $q_1 = 0$, then

$p_1 = 0$, and if $q_1 \neq 0$, then $d = -p_1/q_1$, so that $\sqrt{d}$ would be rational, contrary to assumption. Thus from (1.4)

$$a(p + 2a) + q - a^2 + b^2d = 0$$
$$b(p + 2a) = 0\,.$$

In the second equation $b \neq 0$ and so $p + 2a = 0$, $p = -2a$. Hence, the first equation becomes $q - a^2 + b^2d = 0$, or $q = a^2 - b^2d$. This proves our assertions and shows that $2a$ and $a^2 - b^2d$ are both rational integers.

Evidently, $2a$ and $a^2 - b^2d$ are rational integers if a and b are rational integers. However, as we have seen this condition is not necessary. $\beta = \frac{1}{2} - \frac{1}{2}\sqrt{5}$ is an algebraic integer, which belongs to the field $Ra(\sqrt{5})$. For this number, $a = \frac{1}{2}$ and $b = -\frac{1}{2}$ while $d = 5$. But although a and b are not integers, $2a = 1$ and $a^2 - b^2d = \frac{1}{4} - \frac{5}{4} = -1$ are integers, as predicted by our general conclusion.

2. REPRESENTATION OF INTEGERS

Let us investigate in more detail the conditions under which $a + b\sqrt{d}$ is an integer. We know that if $a + b\sqrt{d}$ is an integer, then $2a = -p$ is an integer. If $2a$ is even, then a is itself an integer. In that case, a^2 also is an integer and the same is then true of $b^2d = a^2 - q$. We are going to show that b^2d can be an integer only if b is an integer. For let $b = m/n$ where m and n are rational integers such that $n > 0$ and such that the fraction m/n is reduced. Then $b^2d = m^2d/n^2$ is supposed to be an integer. If $n = 1$ then we have finished, for then $b = m$ is an integer. Suppose now that $n > 1$. Then n is divisible by a positive prime number, r. (n may of course be divisible by several positive primes. We pick one of them.) But $n^2b^2d = m^2d$ so if $r|n$ then $r^2|n^2$, hence $r^2|n^2b^2d$, that is, $r^2|m^2d$. Now since m/n is reduced, r cannot divide m, hence r cannot divide m^2. Hence $r^2|d$. But this is impossible since d was supposed to be square-free. This shows that n cannot be greater than 1, and so b must be an integer.

Suppose now that $2a = -p$ is odd so that a is *half an odd integer.* It is convenient to consider three distinct cases, according to the class, A_1, A_2, or A_3, to which d belongs modulo 4. That is to say,

(i) in the first case, $d \equiv 1(4)$
(ii) in the second case, $d \equiv 2(4)$
(iii) in the third case, $d \equiv 3(4)$.

The possibility that $d \equiv 0(4)$ can be discounted since in that case d would be divisible by 4 and so d would not be square-free, contrary to assumption.

Consider the second of the three cases enumerated above, $d \equiv 2(4)$. In that case we are going to show that if a is half an odd integer, $q = a^2 - b^2d = (p^2/4) - b^2d$ cannot be an integer for *any* choice of the rational number b. For let $b = m/n$ where m and n are rational integers and m/n is reduced, and let $n = 2^kt$ where $k \geq 0$ and t is odd. Since $d \equiv 2(4)$, it is of the form $d = 2d' + 2 = 2(d' + 1)$ where $d' + 1$ is odd since d is square-free. Hence

$$q = a^2 - b^2d = \frac{p^2}{4} - \frac{2m^2(d' + 1)}{2^{2k}t^2}\,. \tag{2.1}$$

Now, if $k = 0$, the equation

$$p^2t^2 = 4t^2q + 8m^2(d' + 1)\,,$$

which is a consequence of (2.1), leads to a contradiction. For the right-hand side of this equation is evidently even while the left-hand side is odd, since it is the product of the squares of two odd numbers.

If $k = 1$, (2.1) entails

$$p^2t^2 = 4t^2q + 2m^2(d' + 1)$$

which leads to a contradiction by the same argument. Thus, it now remains to be shown that the case $k \geq 2$ cannot be realized either. In that case, we multiply (2.1) by $2^{2k-1}t^2$ and obtain after rearrangement

$$m^2(d' + 1) = 2^{2k-3}p^2t^2 - 2^{2k-1}qt^2\,.$$

Here again the right-hand side must be divisible by 2 while the left-hand side is odd. Thus, the assumption that a is half an odd integer leads to a contradiction in case (ii).

Next, consider case (i), $d \equiv 1(4)$. In this case, $d = 4d' + 1$. If a is half an odd integer, we now have $q = (p^2/4) - b^2(4d' + 1)$. Since q is supposed to be an integer, b cannot be an integer, for this would make $p^2/4 = q + b^2(4d' + 1)$ an integer; that is, $p/2$ would be an integer, contrary to assumption. On the other hand, putting $2b = r$, we find that $q = (p^2/4) - (r^2/4)(4d' + 1)$ and so $r^2(4d' + 1) = p^2 - 4q$. This shows that $r^2(4d' + 1)$ is an integer. Taking into account that $d = 4d' + 1$ is square-free, we conclude that r^2 and hence r are integers. (Why?) Thus, we have shown that b must be half an odd integer, $b = r/2$.

Conversely, if $b = r/2$ where r is odd, we may write

$$q = \frac{p^2}{4} - \frac{r^2}{4}(4d + 1) = \frac{p^2 - r^2}{4} - r^2d' = \frac{p-r}{2} \cdot \frac{p+r}{2} - r^2d' .$$

But $p - r$ and $p + r$ are even since p and r are both odd. It follows that $(p^2 - q^2)/4$ is an integer and hence that q is an integer. Accordingly, in this case β is an integer if both a and b are half odd integers (and also if both a and b are integers).

Finally, consider case (iii), $d \equiv 3(4)$. In this case $d = 4d' + 3 = 4d'' - 1$ where $d'' = d' + 1$. Operating as before we find that if a is half an odd integer then b also must be half an odd integer. And if b is half an odd integer, $b = r/2$, then

$$q = \frac{p^2}{4} - \frac{r^2}{4}(4d'' - 1) = \frac{p^2 + r^2}{4} - r^2d'' .$$

It follows that q is an integer only if $(p^2 + r^2)/4$ is an integer. This, however, is impossible. For p is odd, hence it is either congruent to 1 modulo 4 or it is congruent to 3 modulo 4. (Why?) In the former case p can be written in the form $p = 4p' + 1$, in the latter case as $p = 4p' - 1$. Then, in the first case $p^2 = 16p'^2 + 8p' + 1$. In the second case $p^2 = 16p'^2 - 8p' + 1$. In either case we then have $p^2 = 4p'' + 1$, and similarly $r^2 = 4r'' + 1$, where p'' and r'' are integers. Accordingly, $(p^2 + r^2)/4 = p'' + r'' + \frac{1}{2}$ so that $(p^2 + r^2)/4$ cannot be an integer. Thus, in case (iii), β is an integer only if a and b are both integers.

To sum up, we have shown that in all cases, $\beta = a + b\sqrt{d}$ is an integer if a and b are both integers. Only, if $d \equiv 1(4)$, β can be an integer also for other values of a and b. In that case, β is in fact an integer also if both a and b are half odd integers.

3. RINGS OF INTEGERS

We shall now prove that for $Ra(\alpha) = Ra(\sqrt{d})$ as before, the set of integers in $Ra(\alpha)$, which is denoted by $J(\alpha)$, is a ring and even an integral domain. For this purpose we have to show in particular that the sum and product of elements of $J(\alpha)$ belong to $J(\alpha)$. Let $\beta = a + b\sqrt{d}$ and $\beta' = a' + b'\sqrt{d}$ be two integers in $Ra(\alpha)$, that is, elements of $J(\alpha)$. We wish to prove that $\gamma = \beta + \beta' = (a + a') + (b + b')\sqrt{d}$ is an integer.

Consider the equation

$$(x - [(a + a') + (b + b')\sqrt{d}])(x - [(a + a') - (b + b')\sqrt{d})]) = 0 .$$

You will notice that if γ is substituted on the left-hand side of this equation, then the first factor, $x - [(a + a') + (b + b')\sqrt{d}]$ vanishes, and so the entire left-hand side vanishes. Multiplying the left-hand side out, we find that γ is a root of the equation

$$[x - (a + a')]^2 - (b + b')^2 d = 0,$$

or, equivalently, of

$$x^2 - 2(a + a')x + (a + a')^2 - (b + b')^2 d = 0.$$

Evidently, $2(a + a') = 2a + 2a'$ is a rational integer, bearing in mind that $2a$ and $2a'$ are both rational integers. Thus, in order to make sure that γ is an integer, we only have to show that $(a + a')^2 - (b + b')^2 d$ is an integer. Now

$$(a + a')^2 - (b + b')^2 d = (a^2 - b^2 d) + (a'^2 - b'^2 d) + 2(aa' + bb'd)$$

and on the right-hand side of this equation, $a^2 - b^2 d$ and $a'^2 - b'^2 d$ are certainly integers. There remains to be considered the expression $2(aa' + bb'd)$. We know that a and b and $a' + b'$, respectively, are either simultaneously integers, or they are simultaneously half odd integers. If only a and b, or only a' and b', or neither of these are half odd integers, $2(aa' + bb'd)$ is clearly an integer and we have finished. a, b, a', b' can be half odd integers only if $d \equiv 1(4)$, that is, if $d = 4d' + 1$. Now if $a = p/2$, $b = r/2$, $a' = p'/2$, $b' = r'/2$ where p, r, p', r' are all odd rational integers then

$$2(aa' + bb'd) = \frac{pp' + rr'}{2} + 2rr'd'.$$

But pp' and rr' are odd and so their sum must be even. This shows that $2(aa' + bb'd)$ is a rational integer and proves our assertion that the sum of two integers is an integer.

Next, we consider the product of β and β' as given above. This is

$$\gamma = \beta\beta' = aa' + bb'd + (ab' + a'b)\sqrt{d}.$$

γ now satisfies the equation

$$[x - (aa' + bb'd) - (ab' + a'b)\sqrt{d}]\,[x - (aa' + bb'd)] + [(ab' + a'b)\sqrt{d}] = 0.$$

Expanding the left-hand side

$$x^2 - 2(aa' + bb'd)x + (aa' + bb'd)^2 - (ab' + a'b)^2 d = 0.$$

But $2(aa' + bb'd)$ has already been proved to be an integer. It remains to be shown that the same is true of

$$(aa' + bb'd)^2 - (ab' + a'b)^2 d .$$

Opening the brackets in this expression and rearranging we see that this expression is equal to

$$(a^2 - b^2 d)(a'^2 - b'^2 d) ,$$

which is a rational integer since it is the product of two rational integers.

It is evident that 0 and 1 belong to $J(\alpha)$. (Why?) And, if $\beta = a + b\sqrt{d}$ belongs to $J(\alpha)$ then $-\beta = -a - b\sqrt{d}$ also belongs to $J(\alpha)$. For if β is a root of the quadratic equation with rational integer coefficients $x^2 + px + q = 0$ then $-\beta$ is a root of the equation $x^2 - px + q = 0$, whose coefficients are again rational integers.

The results obtained so far in this section show that the third, fourth, fifth, sixth, and ninth rules enumerated in Chapter 1 are satisfied by $J(\alpha)$, where 0 and 1 are the neutral elements with respect to addition and multiplication, as usual. As for the remaining rules which apply to an integral domain, it is easy to verify that they are satisfied by $J(\alpha)$ because they are satisfied by $Ra(\alpha)$. For example, $\beta + \gamma = \gamma + \beta$ (first rule) is satisfied by all β, γ in $J(\alpha)$ since this is true in the more comprehensive system $Ra(\alpha)$. In this way, we arrive at the conclusion that $J(\alpha)$ is an integral domain (compare condition (i) in section 1 of this chapter).

To conclude this section we observe that every element of $Ra(\alpha)$ is the quotient of two numbers that belong to $J(\alpha)$. (Compare condition (ii) in section 1 of this chapter.) More precisely, if γ belongs to $Ra(\alpha)$ then we may write $\gamma = \beta/n$ where β belongs to $J(\alpha)$ and n belongs to J, which is a subset of $J(\alpha)$.

To see this, let γ be an arbitrary number in $Ra(\alpha)$. Then $\gamma = a + b\alpha$ where a and b are rational. Let n be a common denominator of a and b in the usual sense. That is to say, $na = a'$ and $nb = b'$ are rational integers. Then $\beta = n\gamma = na + nb\alpha = a' + b'\alpha$ is an algebraic integer and belongs to $J(\alpha)$. Hence $\gamma = \beta/n$, the required expression for γ.

4. UNITS

Consider a field $Ra(\alpha) = Ra(\sqrt{d})$ as before (d a rational integer such that $\alpha = \sqrt{d}$ is irrational), and let $J(\alpha)$ be the ring of integers in $Ra(\alpha)$. Under what conditions is an element ε of $J(\alpha)$ a unit? Evidently ε is not a unit if $\varepsilon = 0$ and we discard this case.

4. UNITS

You will recall that ε is a unit in $J(\alpha)$ if there exists a number ε' in $J(\alpha)$ such that $\varepsilon\varepsilon' = 1$. Now let $\varepsilon = a + b\sqrt{d}$ where a and b are rational numbers. The condition for ε to be a unit is that there exist rational numbers x and y such that $\varepsilon' = x + y\sqrt{d}$ is an integer which satisfies $\varepsilon\varepsilon' = 1$, in other words such that

$$(a + b\sqrt{d})(x + y\sqrt{d}) = 1\,,$$

or equivalently,

$$(ax + bdy) + (bx + ay)\sqrt{d} = 1\,.$$

By section 2 of Chapter III, this will be the case if and only if

$$\begin{aligned} ax + bdy &= 1 \\ bx + ay &= 0\,. \end{aligned} \tag{4.1}$$

We are thus faced with the problem of finding out whether (4.1) has a solution x, y such that $x + y\sqrt{d}$ is an integer. We first solve (4.1) in the usual way. That is to say, we multiply the first equation by b and the second equation by a and subtract the first from the second equation. The result is

$$(a^2 - b^2d)y = -b\,.$$

Now $a^2 - b^2d$ must be different from 0, for $a^2 - b^2d = 0$ implies that $(a/b)^2 = d$, so that $\sqrt{d}$ would be rational. $b = 0$ cannot occur for then $a = 0$ and hence $\varepsilon = 0$. Thus

$$y = -\frac{b}{a^2 - b^2d}\,. \tag{4.2}$$

If $b \neq 0$, we may then use the second equation of (4.1) to obtain

$$x = \frac{a}{a^2 - b^2d}\,, \tag{4.3}$$

and we may then verify directly that these values for x and y do indeed satisfy (4.1), If $b = 0$, then $a \neq 0$, so the equation (4.1) is solved, in a unique fashion by $y = 0$, $x = 1/a$, which agree with (4.2) and (4.3).

If you are familiar with determinants then you may solve the system (4.1) more directly. The determinant of (4.1) is $\Delta = a^2 - b^2d$. Having established, as above, that $\Delta \neq 0$, we may write down the solution of (4.1) immediately as

$$x = \frac{\begin{vmatrix} 1 & bd \\ 0 & a \end{vmatrix}}{\Delta} = \frac{a}{a^2 - b^2d}\,, \quad y = \frac{\begin{vmatrix} a & 1 \\ b & 0 \end{vmatrix}}{\Delta} = -\frac{b}{a^2 - b^2d}$$

which agrees with (4.3) and (4.2).

These formulae give us a direct method for deciding whether a given β is a unit. Supposing first that d is not congruent to 1 modulo 4, we know that $\beta = a + b\sqrt{d}$ is an integer if and only if a and b are rational integers. β is a unit if and only if, at the same time, x and y as given by (4.3) and (4.2) are rational integers. At any rate, this will be the case if $a = 1$, $b = 0$, $\beta = 1$, for then $x = 1$, $y = 0$; and also if $a = -1$, $b = 0$, $\beta = -1$ for then $x = -1$, $y = 0$.

The situation is a little more complicated if $d \equiv 1(4)$. In that case β is an integer and a unit under the conditions just stated, but also if a and b are integers and x, y are half odd integers, and if a and b are half odd integers and x, y are either both integers or both half odd integers. All these possibilities can be checked by means of (4.3) and (4.2). Here again, 1 and -1 are units as can also be seen directly. (Why?)

However, this procedure does not enable us to get a clear view of the entire set of units in a given quadratic field. Some important properties of this set can be obtained by other means. Thus, we observe that if ε_1 and ε_2 are units in $Ra(\alpha)$, then their product, $\varepsilon_1\varepsilon_2$, is a unit. For the assumption is that ε_1, ε_2, ε_1^{-1}, ε_2^{-1} all belong to $J(\alpha)$. But $J(\alpha)$ is a ring and so $\varepsilon_2^{-1}\varepsilon_1^{-1} = (\varepsilon_1\varepsilon_2)^{-1}$ also belongs to $J(\alpha)$. This shows that $\varepsilon_1\varepsilon_2$ is a unit. Even more simply, if ε is a unit in $J(\alpha)$, then $\varepsilon^{-1} = 1/\varepsilon$ is a unit.

You will now have no difficulty in verifying that the units of $J(\alpha)$ constitute a commutative group with respect to the operation of multiplication. That is to say, if in rules one to five of Chapter I, section 2, we replace addition by multiplication, then the rules are satisfied by the set of units in $J(\alpha)$. Notice that the neutral element with respect to the group operation is 1 and not 0, which does not even belong to the set. (Instead of talking of replacing addition by multiplication in the rules, it would be more correct to say that we *interpret* the word "addition" in the rules by the operation of multiplication in the set of units.)

For any number $\beta = a + b\sqrt{d}$ in $Ra(\alpha) = Ra(\sqrt{d})$, we call the number $a - b\sqrt{d}$ the *conjugate* of β and we denote it by $\bar{\beta}$. If β is an integer then $\bar{\beta}$ is an integer. (Why?) We may regard the bar on β(in $\bar{\beta}$) as a symbol for an *operation* on the element of $Ra(\alpha)$ which maps $Ra(\alpha)$ on itself. Then $\bar{\bar{\beta}} = \beta$ (Why?) and $(\overline{-\beta}) = -\bar{\beta}$, and for any β_1 and β_2,

$$\overline{\beta_1 + \beta_2} = \bar{\beta}_1 + \bar{\beta}_2, \quad \overline{\beta_1\beta_2} = \bar{\beta}_1\bar{\beta}_2 . \quad \text{(Why?)}$$

The number

$$\beta\bar{\beta} = (a + b\sqrt{d})(a - b\sqrt{d}) = a^2 - b^2 d ,$$

which we have encountered previously, is called the *norm* of β, write $N(\beta)$. Evidently $N(\beta)$ is a rational number. As we have shown (section 1 of this chapter) it is a rational integer if β is an integer. Then $N(\beta) = N(\bar{\beta}) = \beta\bar{\beta}$. If β_1 and β_2 are any two numbers in $Ra(\alpha)$ then

$$N(\beta_1)N(\beta_2) = \beta_1\bar{\beta}_1\beta_2\bar{\beta}_2 = \beta_1\beta_2\bar{\beta}_1\bar{\beta}_2 = \beta_1\beta_2\overline{\beta_1\beta_2} = N(\beta_1\beta_2)\,.$$

You may have thought for a moment that our definition of $N(\beta)$ was ambiguous since it depends on the way in which we write $\beta = a + b\sqrt{d}$ with rational a and b. However, since we showed earlier that there is only one way of writing β in this form, our definition is a good one.

If β^{-1} denotes the reciprocal of β, as usual, then $\beta\beta^{-1} = 1$, and $N(\beta)N(\beta^{-1}) = N(1)$. But $1 = 1 + 0\sqrt{d}$ and so $N(1) = (1 + 0\sqrt{d})(1 - 0\sqrt{d}) = 1$. Hence

$$N(\beta)N(\beta^{-1}) = 1\,.$$

Suppose now that β is a unit. In that case both β and β^{-1} are integers and so $N(\beta)$ and $N(\beta^{-1})$ are rational integers. This shows that $N(\beta)$ is a unit in the ring of rational integers (since its reciprocal is a rational integer). But the only unit in the ring of rational integers, J, are 1 and -1 and so we have shown

If β is a unit then $N(\beta) = 1$ or $N(\beta) = -1$.

On the other hand if $N(\beta) = a^2 - b^2d = \pm 1$ then (see (4.3) and (4.2)) either $x = a$ and $y = -b$, or $x = -a$ and $y = b$. It follows that if a and b are both rational integers then x and y are both rational integers, and if a and b are both half odd integers then x and y are both half odd integers. We conclude that if $N(\beta) = 1$ or $N(\beta) = -1$ and β is an integer, then β^{-1} is an integer and so β is a unit. Thus, we have proved also the converse of the previous result.

If β is an integer and $N(\beta) = 1$ or $N(\beta) = -1$ then β is a unit.

This theorem can be established also by another method. As we argued previously, $\beta = a + b\sqrt{d}$ is a root of the polynomial equation

$$[x-(a + b\sqrt{d})]\;[x-(a - b\sqrt{d})] = 0\,,$$

or, which is the same,

$$x^2 - 2ax + a^2 - b^2d = 0\,,$$

that is,

$$x^2 - 2ax + N(\beta) = 0\,.$$

Thus if β is an integer and $N(\beta) = \pm 1$ then

$$\beta^2 - 2a\beta \pm 1 = 0$$

where $2a$ is a rational integer. Putting $\gamma = 1/\beta = \beta^{-1}$, $\beta = 1/\gamma$, and substituting $1/\gamma$ in the last equation for β, we obtain

$$\frac{1}{\gamma^2} - 2a\frac{1}{\gamma} \pm 1 = 0$$

or, which is the same,

$$\gamma^2 \mp 2a\gamma \pm 1 = 0 .$$

In this connection $\pm$ means "either plus or minus" and the juxtaposition of $\mp$ and $\pm$ means that either the sign prefixed to $2a\gamma$ is $-$, in which case the constant term is $+1$, or the sign prefixed to $2a$ is $+$, in which case the constant term is -1. In either case, γ is a root of the quadratic equation with integer coefficients

$$x^2 \mp 2ax \pm 1 = 0$$

and so γ is an integer. But γ is the reciprocal of β and so β is a unit.

5. UNITS IN $J(\sqrt{2})$

We shall now show in the particular case for which $d = 2$, $\alpha = \sqrt{2}$, how to determine all units in the ring of integers $J(\alpha) = J(\sqrt{2})$. For this value of d, we have $d \equiv 2(4)$ so that $\beta = a + b\sqrt{2}$ is an integer if and only if a and b are rational integers. In order that β may be a unit a and b must also satisfy the condition

$$a^2 - 2b^2 = 1$$

or else the condition

$$a^2 - 2b^2 = -1 .$$

Consider the latter possibility. You will see immediately that it is satisfied by $a = 1$, $b = 1$. Thus, the number $\varepsilon_0 = 1 + \sqrt{2}$ is a unit. But we have shown that the set of units constitutes a group with respect to multiplication, and so

$$\epsilon_0^{-1} = \frac{1}{1+\sqrt{2}} = \frac{1-\sqrt{2}}{(1+\sqrt{2})(1-\sqrt{2})} = \frac{1-\sqrt{2}}{1-2} = -1+\sqrt{2}$$

is a unit. For the same reason, all positive powers of ε_0, that is, the numbers $(1 + \sqrt{2})^2$, $(1 + \sqrt{2})^3$, $(1 + \sqrt{2})^4$, . . . , are units, and similarly $(-1 + \sqrt{2})^2$, $(-1 + \sqrt{2})^3$, . . . , are all units. But 1 and -1 are known to be units, and so the product of any of the units obtained so far by -1 is a unit. Observe that 1 itself may be regarded as the zeroth power of $1 + \sqrt{2}$, and the positive powers of $-1 + \sqrt{2}$ are the corresponding negative powers of $1 + \sqrt{2}$; that is, $(1 + \sqrt{2})^{-n} = (-1 + \sqrt{2})^n$, $n = 1, 2, 3, \ldots$. Summing up, we have found that the following are units in $J(\sqrt{2})$.

$$\pm 1, \pm(1 + \sqrt{2}), \pm(1 + \sqrt{2})^2, \pm(1 + \sqrt{2})^3, \ldots \tag{5.1}$$
$$\pm(-1 + \sqrt{2}), \pm(-1 + \sqrt{2})^2, \pm(-1 + \sqrt{2})^3, \ldots .$$

We now intend to prove that *all* units of $J(\sqrt{2})$ are listed under (5.1). First of all, $\varepsilon_0 = 1 + \sqrt{2}$ is greater than 1, more particularly $2.41 < \varepsilon_0 < 2.42$ since $1.41 < \sqrt{2} < 1.42$. It follows that $\varepsilon_0^{-1} = -1 + \sqrt{2}$ is smaller than 1 though positive, more precisely, $0.41 < \varepsilon_0^{-1} < 0.42$.

You may be inclined to accept the assertion that if a real number is greater than 1, then its powers increase beyond all bounds, and this is indeed true. Here we shall require only the more particular fact that if a number γ is greater than 2 and if M is an arbitrary positive number, then there exists a natural number n such that $\gamma^n > M$. Now if $\gamma > 2$ then $\gamma^n > 2^n$ and so it is sufficient to prove that for any given real M we can find a natural number n such that $2^n > M$.

Now if there are any *real* numbers M for which the assertion is not true then there must also exist a natural number of this kind. (Why?) If so, there exists a smallest natural number, M_0, such that $2^n \leq M_0$ for all natural numbers n. Now $M_0 > 1$ since $2 = 2^1 > 0$ and $2^1 > 1$. It follows that $M_0 - 1$ is a positive natural number. Since it is smaller than M_0, there must exist a natural number n such that $2^n > M_0 - 1$, and hence $2^n + 1 > M_0$. Now $2^{n+1} \geq 2^n + 1$ for all natural numbers n, since $2^{n+1} - 2^n = 2^n$ is non-negative (zero or positive) for all natural numbers n. Hence, $2^{n+1} > M_0$, which is contrary to the defining property of M_0. This proves our assertion.

We are going to show that there cannot exist a unit ε in $J(\sqrt{2})$ such that $1 < \varepsilon < \varepsilon_0 = 1 + \sqrt{2}$. For let $\varepsilon = a + b\sqrt{2}$, where ε is a unit, then $a^2 - 2b^2 = 1$, or $a^2 - 2b^2 = -1$, as was shown previously. Suppose first that $a^2 - 2b^2 = 1$. Dividing this equation by the number $\varepsilon = a + b\sqrt{2}$ which is positive by assumption, we obtain

$$a - b\sqrt{2} = \frac{1}{a + b\sqrt{2}} \cdot$$

The denominator on the right-hand side is greater than 1. Hence

$$0 < a - b\sqrt{2} < 1 . \tag{5.2}$$

At the same time

$$1 < a + b\sqrt{2} < 1 + \sqrt{2}, \text{ by assumption.} \tag{5.3}$$

We may add the chains of inequalities (5.2) and (5.3) (Why?) and obtain

$$1 < 2a < 2 + \sqrt{2} < 4$$

and hence, $0 < a < 2$. Since a is a rational integer we conclude that $a = 1$. Substituting 1 for a in (5.3), we obtain $1 < 1 + b\sqrt{2} < 1 + \sqrt{2}$, and hence $0 < b\sqrt{2} < \sqrt{2}$, and hence $0 < b < 1$. But there is no rational integer between 0 and 1. This shows that the possibility that $a^2 - 2b^2 = 1$ can be ruled out.

Suppose now that $a^2 - 2b^2 = -1$. Then

$$0 > a - b\sqrt{2} = \frac{-1}{a + b\sqrt{2}} > -1 \cdot \tag{5.4}$$

With respect to the second inequality in (5.4), observe that if we divide a negative number (in this case, -1) by a number which is greater than 1 (in this case by $a + b\sqrt{2}$) then the result is *greater* than the former number (the numerator -1).

At the same time, (5.3) shows that

$$-1 > -a - b\sqrt{2} > -1 - \sqrt{2}\,. \tag{5.5}$$

Adding the first two inequalities in (5.4) and (5.5), we obtain $-1 > -2b\sqrt{2}$. This shows that b is positive. (Why?) Adding the second inequalities in (5.4) and (5.5) we then deduce that $-2b\sqrt{2} > -2 - \sqrt{2}$ and hence that $2b\sqrt{2} < 2 + \sqrt{2}$. (Why?) Hence

$$b < \frac{2 + \sqrt{2}}{2\sqrt{2}} = \frac{1}{2}(\sqrt{2} + 1) < 1.3,$$

and so b must be equal to 1. Substituting 1 for b in (5.5)

$$-1 > -a - \sqrt{2} > -1 - \sqrt{2}$$

and hence

$$1 < a + \sqrt{2} < 1 + \sqrt{2}\,. \tag{5.6}$$

The second inequality of (5.6) shows that $a < 1$. The first inequality shows that $a > 1 - \sqrt{2} > -0.5$. Thus $a = 0$. But this would imply that $a^2 - 2b^2 = -2b^2 = -1$, $2b^2 = 1$, and there is no rational integer b which satisfies this condition. This rules out the alternative $a^2 - 2b^2 = -1$. Accordingly, there is no unit ε in $J(\sqrt{2})$ such that $1 < \varepsilon < \varepsilon_0$.

Next we show that there can be no unit ε in the open interval between ε_0^n and ε_0^{n+1} (that is, it is impossible that $\varepsilon_0^n < \varepsilon < \varepsilon_0^{n+1}$) for $n = 1, 2, \ldots$. For if

$$\varepsilon_0^n < \varepsilon < \varepsilon_0^{n+1},\ n = 1, 2, \ldots,$$

then, multiplying by ε_0^{-n},

$$1 < \varepsilon\varepsilon_0^{-n} < \varepsilon_0 . \tag{5.7}$$

But ε_0^{-n} is a unit, and the units form a group with respect to multiplication. Hence $\varepsilon\varepsilon_0^{-n}$ is a unit in the open interval between 1 and ε_0. However, we have already shown that no such unit exists. Thus, there can be no unit in $J(\sqrt{2})$ between ε_0^n and ε_0^{n+1}.

So far we have shown that the only units in $J(\sqrt{2})$ which are greater than 1 are $\varepsilon_0, \varepsilon_0^2, \varepsilon_0^3, \ldots$.

Now let ε be a unit in $J(\sqrt{2})$ which is greater than 0 but smaller than 1. Then ε^{-1} is a unit which is greater than 1. Accordingly, $\varepsilon^{-1} = \varepsilon_0^n$ for some positive natural number n, and so $\varepsilon = \varepsilon_0^{-n}$. Thus ε is included under (5.1).

Finally, if ε is a unit in $J(\sqrt{2})$ and is negative, then $-\varepsilon$ is a positive unit in $J(\sqrt{2})$; for since ε^{-1} is an integer, $-\varepsilon^{-1} = (-\varepsilon)^{-1}$ must be an integer as well. Thus, $-\varepsilon$ is one of two numbers $1, \varepsilon_0, \varepsilon_0^2, \varepsilon_0^3, \ldots$, or $\epsilon_0^{-1}, \varepsilon_0^{-2}, \varepsilon_0^{-3}, \ldots$, and so ε is again included under (5.1). Accordingly, we have proved that all units of $J(\sqrt{2})$ are contained in the (infinite) list (5.1).

In other fields, the situation may be more complicated. For example, in $Ra(\sqrt{10})$ the units $\varepsilon = a + b\sqrt{10}$ have to satisfy the condition

$$a^2 - 10b^2 = \pm 1$$

This condition is satisfied by ± 1, and also by $\varepsilon_0 = 3 + \sqrt{10}$ and its positive and negative powers, and their inverses (that is, by $\pm\varepsilon_0^n$, where n is any rational integer). However, it can be shown that in this case there are infinitely many additional units.

EXERCISES

1. Let α be a root of the equation $x^2 + x + 1 = 0$. Show that α is a unit in $J(\alpha)$.

2. Find all units in $J(\sqrt{-7})$.

3. Find all units in $J(\sqrt{-3})$.

4. Suppose that ε is a unit in $J(\alpha)$. Show that its conjugate, $\bar{\varepsilon}$, also is a unit.

5. Let R_1 and R_2 be two distinct quadratic fields. Show that they have only the rational numbers in common.

V

Primes and Factorization in Quadratic Number Fields

1. PRIMES

We recall that the number β is said to be prime in a given integral domain if β is neither zero nor a unit and if every divisor of β either is a unit or else is a number associated with β.

To test if a number $\beta = a + b\sqrt{d}$, $d \not\equiv 1(4)$ is prime in a given integral domain, $J\ (\sqrt{d})$, we proceed as follows. We first check that β is different from zero and that it is not a unit. The latter condition (that is, β *not* being a unit) is satisfied if $N(\beta) = a^2 - b^2d$ is equal neither to 1 nor to -1. Now suppose that $\beta = \beta_1\beta_2$ where β_1 and β_2 also belong to $J(\sqrt{d})$. Then

$$\beta_1 = a_1 + b_1\sqrt{d} \qquad \beta_2 = a_2 + b_2\sqrt{d}\ ,$$

and

$$N(\beta_1) = a_1^2 - b_1^2d \qquad N(\beta_2) = a_2^2 - b_2^2d\ ,$$

and, as we know from Chapter IV, section 5, $N(\beta) = N(\beta_1)N(\beta_2)$. Thus,

$$a^2 - b^2d = (a_1^2 - b_1^2d)(a_2^2 - b_2^2d)\ . \tag{1.1}$$

If, for the given $\beta = a + b\sqrt{d}$, (5.1) is satisfied only by rational integers, a_1, b_1, a_2, b_2 such that $a_1^2 - b_1^2d = \pm 1$ or $a_2^2 - b_2^2d = \pm 1$, then either β_1 or β_2 is a unit and so the second factor, β_2 or β_1, is associated with β. It follows that β is prime.

Consider the following examples.

(i) Let $d = 2$, so that we are dealing with $J(\sqrt{2})$. Then $d \equiv 2(4)$ so that the procedure outlined above is applicable. Let $\beta = \sqrt{2}$, so that

$a = 0$, $b = 1$, (that is, $\beta = 0 + 1\sqrt{d}$). Then $N(\sqrt{2}) = -2$, so that $\sqrt{2}$ is not a unit. Equation (1.1) becomes

$$-2 = (a_1^2 - 2b_1^2)(a_2^2 - 2b_2^2)\,, \tag{1.2}$$

so we have only the following possibilities.

$$a_1^2 - 2b_1^2 = 1 \qquad a_2^2 - 2b_2^2 = -2$$

or

$$a_1^2 - 2b_1^2 = -1 \quad a_2^2 - 2b_2^2 = 2$$

or, two more possibilities, obtained by interchanging the subscripts 1 and 2. In the first two cases β_1 is a unit, so β_2 is associated with β, in the remaining two cases β_2 is a unit and β_1 is associated with β. Thus $\beta = \sqrt{2}$ is prime in $J(\sqrt{2})$.

On the other hand, the number 2 is not prime in $J(\sqrt{2})$ since $2 = \sqrt{2} \cdot \sqrt{2}$, although 2 is prime in J. This shows that a number is or is not prime only in relation to a specified integral domain.

(ii) Let $d = 10$, $d \equiv 2(4)$. The integral domain under consideration is $J(\sqrt{10})$. An example of a rational integer which is prime in J but not in $J(\sqrt{10})$ is 41. Evidently 41 is prime in J. (Why?) However, in $J(\sqrt{10})$ we have

$$41 = (9 + 2\sqrt{10})(9 - 2\sqrt{10})$$

and $N(9 + 2\sqrt{10}) = N(9 - 2\sqrt{10}) = 41$ so that 41 is the product of two numbers in $J(\sqrt{10})$ neither of which is a unit. Thus 41 has divisors which are neither units nor associated with 41. 41 is *not* a prime. ($9 + 2\sqrt{10}$ and $9 - 2\sqrt{10}$ are not associated with 41. Why?)

We shall now give four examples of primes in $J(\sqrt{10})$.

The number 2 is a prime in $J(\sqrt{10})$.

In this case, $N(\beta) = N(2) = N(2 + 0\sqrt{10}) = 4$, so (5.1) becomes

$$4 = (a_1^2 - 10b_1^2)(a_2^2 - 10b_2^2)\,.$$

Unless either β_1 or β_2 is a unit (and these cases are ruled out) we must have either

$$a_1^2 - 10b_1^2 = a_2^2 - 10b_2^2 = 2\,, \tag{1.3}$$

or

$$a_1^2 - 10b_1^2 = a_2^2 - 10b_2^2 = -2\,, \tag{1.4}$$

where a_1, b_1, a_2, b_2 are rational integers. However, we shall show that there are no such rational integers. More generally, there are no rational integers x and y such that

$$x^2 - 10y = 2 \quad \text{or} \quad x^2 - 10y = -2 . \tag{1.5}$$

If there are such rational integers, we may suppose x to be a natural number, since $x^2 = (-x)^2$. We now divide x with remainder by 10, so $x = 10q + t$, where $0 \leq t \leq 9$. Then $x^2 = 100q^2 + 20qt + t^2$. Substituting this expression for x^2 in the first equation of (1.5) we obtain, after some slight manipulation,

$$t^2 - 2 = 10(y - 10q^2 - 2qt) \tag{1.6}$$

while the second equation yields

$$t^2 + 2 = 10(y - 10q^2 - 2qt) . \tag{1.7}$$

Now t can only take rational integer values from 0 to 9, so we have the table shown in Figure 6.

t	0	1	2	3	4	5	6	7	8	9
t^2	0	1	4	9	16	25	36	49	64	81
$t^2 - 2$	-2	-1	2	7	14	23	34	47	62	79
$t^2 + 2$	2	3	6	11	18	27	38	51	66	83

Fig. 6.

The table shows that neither the left-hand side of (1.6) nor the left-hand side of (1.7) is divisible by 10 for any of the values $t = 0, 1, \ldots, 9$, although the right-hand sides of these equations clearly *are* multiples of 10. It follows that (1.3) and (1.4) can have no solution in rational integers, since any solution of (1.3) provides a solution of the first equation of (1.5) by means of $x = a_1$, $y = b_1^2$, and any solution of (1.4) provides a solution of the second equation of (1.5) by the same substitution.

The number 3 *is a prime in* $J(\sqrt{10})$.

In this case $N(\beta) = N(3) = 9$, so (5.1) becomes

$$9 = (a_1^2 - 10b_1^2)(a_1^2 - 10b_2^2) .$$

So if neither β_1 nor β_2 is a unit we must have either

$$a_1^2 - 10b_1^2 = a_2^2 - 10b_2^2 = 3 \tag{1.8}$$

or

$$a_1^2 - 10b_1^2 = a_2^2 - 10b_2^2 = -3\,.$$

However, we are going to show that neither

$$x^2 - 10y = 3 \quad \text{nor} \quad x^2 - 10y = -3$$

has a solution in rational integers. For, putting $x = 10\,q + t$, as before, we should then have

$$t^2 - 3 = 10(y - 10q^2 - 2qt) \quad \text{or} \quad t^2 + 3 = 10(y - 10q^2 - 2qt)$$

for some t, $0 \leq t \leq 9$. In that case, $t^2 - 3$ or $t^2 + 3$ would have to be divisible by 10 but the table given in Figure 7 shows that no such t exists.

t	0	1	2	3	4	5	6	7	8	9
$t^2 - 3$	-3	-2	1	6	13	22	33	52	61	84
$t^2 + 3$	3	4	7	12	19	28	39	46	67	78

Fig. 7.

The number $4 + \sqrt{10}$ *is a prime in* $J(\sqrt{10})$.

For, in this case, $N(\beta) = N(4 + \sqrt{10}) = 16 - 10 = 6$, so (5.1) becomes

$$6 = (a_1^2 - 10b_1^2)(a_2^2 - 10b_2^2)\,.$$

Unless β_1 or β_2 is a unit $a_1^2 - 10b_1^2$ must be either ± 2 or ± 3 and we have already seen that none of these four possibilities can occur. This proves our assertion. Similarly,

The number $4 - \sqrt{10}$ *is a prime in* $J(\sqrt{10})$.

2. PRIME FACTORIZATION

We are now going to show that every number in $J(\sqrt{d})$ which is not a unit or zero is a product of primes. Here d is an arbitrary integer such that $\sqrt{d}$ is irrational. As we have seen, however, d may, without loss of generality, be supposed to be square-free. In this connection, we include under *product of primes* the possibility that the number is itself a prime, that is, that the number of elements in the product is 1.

We observe that if β is a prime in $J(\sqrt{d})$ and ε is a unit, then $\varepsilon\beta$

also is a prime. For if β_1 is a divisor of $\varepsilon\beta$ which is not a unit and is not associated with $\varepsilon\beta$, $\beta_1\beta_2 = \varepsilon\beta$, then $\varepsilon^{-1}\beta_1\beta_2 = \beta$, so that $\varepsilon^{-1}\beta_1$ is a divisor of β which is not a unit (otherwise β_1 would be a unit—Why?), and is not associated with β (otherwise β_1 would be associated with $\varepsilon\beta$—Why?).

Let B be the set of elements of $J(\sqrt{d})$ which are different from zero, and are not units and yet cannot be written as products of primes. We shall prove that B is empty. Suppose, contrary to our assertion, that B is not empty. Let C be the set of rational numbers which is defined by

$$C = \{n|n = |N(\beta)|, \beta \text{ in } B\} .$$

That is to say, C consists of all numbers which are absolute values of norms of elements of B . C is not empty. It cannot contain the number 0. For if $\beta = a + b\sqrt{d}$ and $|N(\beta)| = |a^2 - b^2d| = 0$ although $\beta \neq 0$, then $d = (a/b)^2$, so that $\sqrt{d} = \pm a/b$ is rational, and this is contrary to one of our assumptions.

C cannot contain the number 1 either. For if $|N(\beta)| = 1$ $N(\beta) = \pm 1$, so that β is a unit. This is again contrary to our assumptions.

Let n_0 be the smallest number which is contained in C. Since n_0 is not 0 or 1, we have $n_0 \geq 2$. There exists a number β_0 in B such that $n_0 = |N(\beta_0)|$. By the definition of C, β_0 cannot be written as a product of primes and hence cannot be a prime. Thus β_0 would have a divisor β_1 which is neither a unit nor is associated with β_0. Let $\beta_0 = \beta_1\beta_2$. Then β_2 cannot be a unit, otherwise β_1 would be associated with β_0. As we have seen $\beta_0 = \beta_1\beta_2$ entails $N(\beta_0) = N(\beta_1)N(\beta_2)$. Hence $|N(\beta_0)| = |N(\beta_1)|\,|N(\beta_2)|$. Since $|N(\beta_0)|$ is positive, $|N(\beta_1)|$ and $|N(\beta_2)|$ must be positive. And since neither β_1 nor β_2 is a unit, both $|N(\beta_1)|$ and $|N(\beta_2)|$ must be greater than 1. But the product of two natural numbers which are greater than 1 is greater than either one of the two factors. (Why?) Hence,

$$|N(\beta_1)| < |N(\beta_0)|, \quad |N(\beta_2)| < |N(\beta_0)| .$$

This shows that β_1 and β_2 do not belong to B. It now follows from the definition of C that both β_1 and β_2 can be written as products of primes in $J(\sqrt{d})$, thus,

$$\beta_1 = \gamma_1\gamma_2 \cdots \gamma_k, \quad \beta_2 = \gamma_{k+1}\gamma_{k+2} \cdots \gamma_{k+l}, \quad k \geq 1, \quad l \geq 1 .$$

But, if so, then

$$\beta_0 = \beta_1\beta_2 = \gamma_1\gamma_2 \cdots \gamma_k\gamma_{k+1}\gamma_{k+2} \cdots \gamma_{k+l} ,$$

also is a product of primes in $J(\sqrt{d})$. This contradicts the fact that β_0 belongs to C and shows that our assumption that C is not empty is untenable, C must be empty. *Conclusion: Every element of* $J(\sqrt{d})$ *which is neither* 0 *nor a unit is a product of primes in* $J(\sqrt{d})$.

3. A MEMORABLE ERROR

Mathematicians, even good mathematicians, are not infallible. Every now and then they make mistakes. Some mistakes are trivial and best forgotten. There are however, also "interesting" mistakes which occur because some theorem which was known to be true in familiar cases was taken for granted also in some new and unfamiliar case. Such a mistake occurred in the work of the mathematician E. E. Kummer (1810–1893). Kummer, who was one of the first to consider algebraic integers tried to solve a famous problem in number theory by making the *unwarranted* assumption that the *unique* factorization theorem of Chapter II, section 6, is always true in rings of algebraic integers. The rings in question were of a somewhat more general nature than the ones considered by us in this book, but we may make the point already in one of the rings considered here. We are in fact going to give an example which shows that *the unique factorization theorem is not true in* $J(\sqrt{10})$.

Indeed consider the number 6 within $J(\sqrt{10})$. Evidently

$$6 = 2 \cdot 3 = (4 + \sqrt{10})(4 - \sqrt{10})\,. \tag{3.1}$$

We know from the preceding section that the numbers 2, 3, $4 + \sqrt{10}$, $4 - \sqrt{10}$, are all prime in $J(\sqrt{10})$. Hence, if the unique factorization theorem were true in $J(\sqrt{10})$ either $4 + \sqrt{10} \sim 2$ and $4 - \sqrt{10} \sim 3$ or $4 - \sqrt{10} \sim 2$ and $4 + \sqrt{10} \sim 3$. By the definition of the relation of association, $\sim$, that is to say that either

(i) there exist units ε_1 and ε_2 such that $4 + \sqrt{10} = \varepsilon_1 2$ and $4 - \sqrt{10} = \varepsilon_2 3$, or

(ii) there exist units ε_1 and ε_2 such that $4 + \sqrt{10} = \varepsilon_1 3$ and $4 - \sqrt{10} = \varepsilon_2 2$.

In the first case, $\varepsilon_1 = (4 + \sqrt{10})/2 = 2 + \frac{1}{2}\sqrt{10}$, and so $N(\varepsilon_1) = (2 + \frac{1}{2}\sqrt{10})(2 - \frac{1}{2}\sqrt{10}) = 4 - \frac{10}{4} = \frac{3}{2}$, so that ε_1 cannot be a unit (for if ε_1 is a unit $N(\varepsilon_1) = \pm 1$). In the second case $N(\varepsilon_2) = N[(4 - \sqrt{10})/2] = \frac{3}{2}$, so that ε_2 cannot be a unit. Thus both possibilities are ruled out, the factorization of 6 in $J(\sqrt{10})$ is not unique in the sense expressed by the unique factorization theorem.

As mentioned earlier, Kummer took it for granted in a similar case that the factorization into primes is unique. However, when his mistake was pointed out to him he did not despair but instead replaced the notion of a prime number by a new concept in which he succeeded in "saving" the unique factorization theorem. An alternative method, which has the same aim in mind, was formulated subsequently by R. Dedekind (1831–1916). The concept of an *ideal* which was introduced by Dedekind, has far-reaching applications not only in number theory but also in other branches of mathematics and even in theoretical physics. However, these applications are beyond our scope.

4. DIVISIBILITY AND FACTORIZATION IN INTEGRAL DOMAINS

Let D be an arbitrary integral domain and let a be an element of D. We define K as the set of all elements of D which are divisible by a. Thus

(4.1) $K = \{b | \text{There exists an element } q \text{ of } D \text{ such that } b = qa\}$.

It follows from this definition that K possesses the following properties.

(i) If the element b_1 and b_2 belong to K then their sum $b_1 + b_2$ belongs to K.

For, by the condition of K, there exist elements of D, q_1 and q_2, such that $b_1 = q_1a$ and $b_2 = q_2a$. Hence,

$$b_1 + b_2 = q_1a + q_2a = (q_1 + q_2)a .$$

This shows that $b_1 + b_2$ also is divisible by a and belongs to K.

(ii) If b_1 and b_2 belong to K then $b_1 - b_2$ belongs to K.

For if $b_1 = q_1a$ and $b_2 = q_2a$ then $b_1 - b_2 = (q_1 - q_2)a$, so that $b_1 - b_2$ is divisible by a and belongs to K.

(iii) If b belongs to K and r is any element of D then rb belongs to K.

For, by the construction of K there exist aq such that $b = qa$. Hence $rb = rqa$. This shows that rb is divisible by a, rb belongs to K.

Evidently a belongs to K since $a = 1a$, a is divisible by itself. Also, 0 belongs to K as we see either directly, or by (ii), since $0 = b - b$, for any b in K.

Every element a of D *generates* a set K by means of the definition (4.1) above, such that K possesses the properties (i), (ii), and (iii). For example, if $a = 0$, then the corresponding K contains 0 alone (Why?) and is called O, while if $a = 1$ then the corresponding K coincides with the entire D.

We now ask ourselves whether, for every non-empty subset K of D, there exists an element a of D such that K is given by (5.1); that is, such that K consists of all elements of D which are divisible by a. It turns out that the answer to this question depends on D. We are going to prove

Theorem. If D is the ring of rational integers, $D = J$, then for every non-empty subset K of D which satisfies (i), (ii), and (iii) above, there exists an element a of D such that K consists of all elements of D that are divisible by a.

Proof. Let K be a non-empty subset of D which satisfies (i), (ii), and (iii). If K contains only 0, then $a = 0$ satisfies the conclusion of the theorem. If K contains numbers other than 0, then it must contain also positive numbers. For if b is a negative number in K, then the positive number, $-b = 0 - b$, also belongs to K, by (ii). Let a be the smallest positive number which is contained in K. (There is such a number—Why?) We claim that a is a divisor of all elements of K. Indeed, let b be any positive number in K. Dividing b with remainder by a, we obtain

$$b = qa + r$$

where q and r are rational numbers. $0 \leq r < a$. Then $r = b - qa$. Now a belongs to K, so qa belongs to K, by (iii). And since both b and qa belong to K, $r = b - qa$ belongs to K, by (ii). But r is smaller than a and a is the smallest positive natural number which belongs to K. This forces us to conclude that $r = 0$, $b = qa$, so that b is divisible by a.

Suppose next that b is a negative number which is contained in K. Then $-b$ also belongs to K, and $-b$ is positive and hence is divisible by a; $-b = qa$ as proved already. But then $b = -qa = (-q)a$, so that b also is divisible by a. Finally, 0 is divisible by a, for $0 = 0a$.

We have now proved that all elements of K are divisible by a. Conversely, all elements of J which are divisible by a are contained in K, by (iii). This proves that a satisfies the conclusion of the theorem.

The corresponding conclusion, however, is not true for all integral domains. This is a consequence of the second of the following theorems.

Theorem. Suppose that the integral domain D has the property that for every non-empty subset K of D which satisfies conditions (i), (ii), and (iii) above, there exists an element a of D such that K consists of all elements of D that are divisible by a. Suppose also that the prime p in D is a divisor of the product b_1b_2 of two elements of D. Then p is a divisor of b_1 or of b_2 or of both b_1 and b_2.

Proof. Suppose that the assumptions of the theorem are satisfied and consider the set K which consists of all elements b of D that can be written in the form $b = mp + nb_1$ where m and n are arbitrary elements of D. Thus,

$$K = \{b | b = mp + nb_1, m \text{ and } n \text{ in } D\} .$$

Then K satisfies conditions (i), (ii), and (iii). (Why?) K contains both p and b, since $p = 1p + 0b_1$ and $b_1 = 0p + 1b_1$. By one of the assumptions of the theorem, there exists an element a of D such that K consists of all elements of D that are divisible by a. Thus a is a divisor of p and at the same time, is a divisor of b_1. But p is prime so a is either a unit or is associated with p. If a is associated with p then p is a divisor of b_1. For in that case, $a = \varepsilon p$ where ε is a unit. Also, $b_1 = ra$ for some r in D since b_1 belongs to K. Hence, $b_1 = r\varepsilon p$ so that p is a divisor of b_1, proving the theorem in this case.

Suppose now that a is a unit, then a possesses a reciprocal a^{-1} in D, $aa^{-1} = 1$. And since a is in K, (iii) shows that 1 also is in K. Hence, there exists elements m_1 and n_1 of D such that

$$1 = m_1p + n_1b_1 .$$

Multiplying this equation by b_2, we obtain

$$b_2 = m_1b_2p + n_1b_1b_2 .$$

But p is supposed to be a divisor of b_1b_2, so that $b_1b_2 = qp$ for some element q of D. Hence

$$b_2 = m_1b_2p + n_1qp = (m_1b_2 + n_1q)p .$$

This shows that p is a divisor of b_2 and completes the proof of the theorem.

Theorem. Suppose that the integral domain D has the property assumed in the preceding theorem. That is to say, for every non-empty subset K of D which satisfies conditions (i), (ii), and (iii), there exists an element a of D such that K consists of all elements of D that are divisible by a. Suppose that

$$b = p_1p_2 \cdots p_k = \varepsilon p_1'p_2' \cdots p_l', \; k \geq 1, \; l \geq 1 , \tag{4.2}$$

where the p_i, $1 \leq i \leq k$ and p_i', $1 \leq i \leq l$ are all prime and ε is a unit. Then $k = l$, and we may renumber the p_i' in such a way that $p_i \sim p_i'$, $i = 1, \ldots, k$.

Proof. We prove the theorem by induction on the number of elements in the first product, k, beginning with $k = 1$.

By the preceding theorem, p_k is a divisor of p_l', or else of $\varepsilon p_1'$,

$\ldots, p_{l-1}'$ (since p_k is a divisor of the product of $\varepsilon p_1' \ldots p_{l-1}'$ and of p_l'). In the latter case, p_k is a divisor of p_{l-1}', or else of $\varepsilon p_1' \ldots p_{l-2}'$, by the same argument. Continuing in this way, we find that p_k is either a divisor of some p_i' for $i > 1$ or else of $\varepsilon p_1'$, and this is true even if $l = 1$. Now if p_k is a divisor of $\varepsilon p_1'$ then, again by the preceding theorem, it must be a divisor of ε or of p_1'. But if p_k were a divisor of ε, $\varepsilon = qp_k$, then since $\varepsilon\varepsilon^{-1} = 1$, we should have $p_k q \varepsilon^{-1} = 1$, that is to say, p_k would have its reciprocal in D, and would be a unit. But this is impossible, since p is prime. It follows that p_k divides p_1' in this case. Thus, quite generally, p_k divides some p_i', $1 \leq i \leq l$. For this i, p_k must be associated with p_i'. For since p_i' is a prime, its divisors are either units or are associated with it, and p_1 is not a unit. Hence $p_k \sim p_i'$, $p_i' = \varepsilon_1 p_1$, where ε_1 is a unit. Then (4.2) becomes, after a slight rearrangement,

$$p_1 p_2 \cdots p_k = p_k \varepsilon \varepsilon_1 p_1' \cdots p'_{i-1} p'_{i+1} \cdots p_l' .$$

By the cancellation rule (a consequence of the eleventh rule of Chapter I) we may delete p_1 on both sides of this equation and obtain

$$p_1 p_2 \cdots p_{k-1} = \varepsilon \varepsilon_1 p_1' \cdots p'_{i-1} p'_{i+1} \cdots p_l' . \tag{4.3}$$

If $k = 1$, then the expression on the left-hand side of (4.3) has to be replaced by 1. But in that case no prime p_i' can appear on the right-hand side of that equation either. For any such p_i' would then be a divisor of 1, and hence a unit. (Why?) And a prime cannot be a unit, by definition. It follows that if $k = 1$, then $l = 1$, and $p_1 \sim p_1'$.

Suppose now that the assertion of the theorem has been proved for $k - 1 = n \geq 1$. To prove it for k we observe that in (4.3) $\varepsilon\varepsilon_1$ is a unit (being the product of two units). Thus, (4.3) may be regarded as a special case of (4.2) and if the assertion of the theorem has been proved for $k - 1$ it follows that $l - 1 = k - 1$ and $p_1' \ldots p'_{i-1} p'_{i+1} \ldots p_l'$ can be renumbered so that $p_1 \sim p_1'$, $p_2 \sim p_2'$, $\ldots$, $p_{k-1} \sim p'_{k-1}$. Renumbering p_i' so that it becomes the last prime on the right-hand side, we then have $p_k \sim p_k'$, and the theorem is proved.

The important case of the theorem occurs when $\varepsilon = 1$, but other units may occur in the course of the proof for that case. (Why?) The theorem includes the second part of the unique factorization theorem for rational integers, formulated at the end of Chapter II, section 6. On the other hand, since we have given an example which shows that factorization into primes in $J(\sqrt{10})$ is not unique it follows that the assumptions of our last theorem are not satisfied by $J(\sqrt{10})$. Thus, $J(\sqrt{10})$ possesses subsets K which satisfy conditions (i), (ii), and (iii), but for which there is no a such that K consists of all elements of $J(\sqrt{10})$ that are divisible by a. However, not all rings $J(\sqrt{d})$ have this property. For example, the ring $J(\sqrt{2})$ satisfies the assumptions of our last theorem and hence, its conclusion.

EXERCISES

1. Let β be an integer in a ring of integers $J(\alpha)$. Show that if $N(\beta)$ is prime in the ring of rational integers then β is prime in $J(\alpha)$.

2. Show that the numbers 3, 7, and $1 + 2\sqrt{-5}$ are all prime in $J(\sqrt{-5})$.

3. Use the results of Exercise 2 in order to show that the number 21 possesses two distinct factorizations into primes in $J(\sqrt{-5})$ in the sense of Chapter 5, section 3.

VI
Ideals

1. IDEALS

Let D be an integral domain. A non-empty subset K of D is called an *ideal* if it satisfies conditions (i), (ii), and (iii) given at the beginning of section 4 of Chapter V (page 63). A requirement equivalent to (i) and (ii) is that K constitutes a group with respect to addition. (Why?)

Let $a_1, \ldots, a_n$, $n \geq 1$, be a finite set of elements in an integral domain, D. Let K be the set of the elements b of D that can be written in the form $b = r_1a_1 + \ldots + r_na_n$ where $r_1, \ldots, r_n$ are arbitrary elements of D. Then K is an ideal. (Why?) The set $a_1, a_2, \ldots, a_n$ is said to be a basis of K, and we write $K = (a_1, \ldots, a_n)$. In general, an ideal has many bases. If $n = 1$, $K = (a_1)$ then K consists precisely of the elements b of D that are divisible by a_1, $b = r_1a_1$, or, as we shall say also, *that are multiples of* a_1. In this case, K is said to be a *principal ideal.* The assumption of the last theorem of the preceding section is that all ideals in D are principal ideals. Particular principal ideals are $(1) = D$, and $(0) = O$, whose only element is in fact 0.

The decisive point of the following development is that we operate with ideals to some extent as if they were ordinary numbers.

Definition. Let H and K be two ideals in the integral domain D. The *product* of H and K, written $H \cdot K$ or HK, consists of all elements a of D that can be expressed as sums of products.

$$(1.1) \qquad a = a_1b_1 + a_2b_2 + \cdots + a_mb_m, \quad m \geq 1, a_i \text{ in } H, b_i \text{ in } K \text{ for } 1 \leq i \leq m.$$

The product of two ideals is an ideal.

In fact, you will see directly that the sum and difference of two elements which are of the form given by (1.1) are again of the same form.

Now suppose that a is given by (1.1) and so belongs to HK and let r be an arbitrary element of D. Then

$$ra = r(a_1b_1 + a_2b_2 + \cdots + a_mb_m) = (ra_1)b_1 + (ra_2)b_2 + \cdots + (ra_m)b_m$$

where $ra_1, ra_2, \ldots, ra_m$ belong to H. Thus, ra is again of the form of (1.1) and belongs to HK. We conclude that HK is an ideal.

If $H = (a_1, \ldots, a_k)$ *and* $K = (b_1, \ldots, b_l)$, *then* $HK = (a_1b_1, a_1b_2, \ldots, a_2b_2, \ldots, a_kb_l)$.

That is to say, a basis for the product of two ideals with bases $a_1, \ldots, a_k, b_1, \ldots, b_l$ is provided by the set of all products a_ib_j, $1 \leq i \leq k, 1 \leq j \leq l$.

To prove this assertion, we put $M = (a_1b_1, a_1b_2, \ldots, a_2b_1, \ldots, a_kb_l)$—an ideal for which the set of all products a_ib_j, $1 \leq i \leq k, 1 \leq j \leq l$ constitutes a basis—and we show that $HK = M$. We do this by verifying (i) that every element of M is an element of HK, and (ii) that every element of HK is an element of M.

(i) Let a be an element of M. Then we may write a as

$$a = r_{11}a_1b_1 + r_{12}a_1b_2 + \cdots + r_{21}a_2b_1 + \cdots + r_{kl}a_kb_l\,, \tag{1.2}$$

where the r_{ij} are certain elements of D. But $r_{11}a_1, r_{12}a_1, \ldots, r_{21}a_2, \ldots, r_{kl}a_k$, are all elements of H and so the right-hand side of (1.2) is a sum of products of elements of H multiplied by elements of K. This shows that a belongs to HK.

(ii) All elements of HK are sums of products ab where a is in H and b is in K. In order to show that all elements of HK belong to M it is sufficient to verify that any such product is in M; for M is an ideal. Hence, if it contains all such products, it contains also their sums.

Suppose then that a belongs to H and b belongs to K. Then $a = r_1a_1 + \cdots + r_ka_k$ and $b = s_1b_1 + \cdots + s_lb_l$, for certain elements r_i, s_j, of D. Hence,

$$\begin{aligned} ab &= (r_1a_1 + \cdots + r_ka_k)(s_1b_1 + \cdots + s_lb_l) \\ &= r_1s_1a_1b_1 + r_1s_2a_1b_2 + \cdots + r_2s_1a_2b_1 + \cdots + r_ks_la_kb_l\,. \end{aligned}$$

The right-hand side shows that ab can be written as in (1.2). Thus, ab belongs to M. This proves that $HK = M$, as asserted.

In particular, if H and K are principal ideals, $H = (a)$, and $K = (b)$, then $HK = (ab)$. Thus, the product of two principal ideals is a principal ideal, and a basis for the product is provided by the product of the single elements which constitute bases for the two factors.

Let $H = (a)$ *be a principal ideal. Then* $H = D$ *if and only if* a *is a unit.*

For if a is a unit then $1 = a^{-1}a$ where a^{-1} is in D. Let b be any element of D, then $b = ba^{-1}a$. Thus b is in H, $H = D$. Conversely if $H = D$, then 1 belongs to H and so $1 = ra$ for some r in D. This shows that a is a unit, while r is the reciprocal of a.

Since ideals are sets and not individuals, it will save time if we introduce the standard notation used in the algebra of sets. Thus, we shall write $a \in K$ if a is in (is an element of) the set K, and $a \notin K$ if this is not the case. We shall write $H \subset K$ if H is a subset of K, that is, if all elements of H are elements of K. Thus, $H = K$ if and only if $H \subset K$ and at the same time $K \subset H$. The intersection of two sets H and K, that is, the set of elements contained in both H and K, is denoted by $H \cap K$ and, finally, the union of H and K (the set of elements which are contained either in H or in K or in both H and K) is denoted by $H \cup K$. The definition of the product of two ideals H and K shows directly that $HK \subset H \cap K$. (Why?)

For every ideal H, $HD = H$.

Indeed, $HD \subset H \cup D = H$. On the other hand if $a \in H$ then $a \cdot 1 \in HD$ since $1 \in D$. But $a \cdot 1 = a$, so $a \in HD$, every element of H is included in HD, $H \subset HD$. This, together with $HD \subset H$ establishes that $HD = H$.

The fact just proved shows that as far as the multiplication of ideals is concerned, D acts as a neutral element with respect to multiplication, rather like the number 1 in ordinary multiplication. For this reason it is sometimes called the unit ideal and denoted also by I. Evidently, $D = (1) = (\varepsilon)$, where ε is an arbitrary unit.

Similarly, the ideal $O = (0)$ plays a role which is analogous to that of 0 in ordinary multiplication. For $KO = O$ (Why?), just as in ordinary multiplication $a0 = 0$ for any $a \in D$. For this reason (and perhaps also because O contains only 0) O is called the *zero ideal.*

The multiplication of ideals is associative and commutative, $H(KM) = (HK)M$ and $HK = KH$. (Why?)

As with ordinary numbers we may therefore omit the parenthesis in a product of several ideals, $H_1H_2 \cdots H_j$.

We write $H = H^1$, $H \cdot H = H^2$, $H \cdot H \cdot H = H^3$, etc., as for ordinary numbers.

For any a and b in D, $(b) \subset (a)$ if and only if $a|b$.

$a|b$ means, by definition that a divides b, $b = ra$. Now every element c of (b) can be written as $c = sb$. Hence $c = (sr)a$ so that $c \in (a)$. This proves that $(b) \subset (a)$. Conversely, if $(b) \subset (a)$ then $b \in (a)$, hence $b = ra$, that is, $a|b$.

For any a and b in D, $(a) = (b)$ if and only if $a \sim b$.

Suppose $a \sim b$, hence $a|b$ and $b|a$ as proved previously. Then $(b) \subset (a)$ and $(a) \subset (b)$, as shown above, hence $(a) = (b)$. Conversely, if $(a) = (b)$ then $(a) \subset (b)$ and $(b) \subset (a)$, hence $b|a$ and $a|b$, hence $a \sim b$.

Next we introduce the fundamental concept of a *prime ideal.* We are going to give two definitions of this concept, and shall prove immediately that the two definitions are equivalent.

Prime Ideals, First Definition. An ideal P in D is *prime* if for all ideals H and K in D, $HK \subset P$ implies that at least one of the two ideals H or K is a subset of P, that is, $H \subset P$ or $K \subset P$.

Prime Ideals, Second Definition. An ideal P in D is *prime* if for all elements a and b of D, $ab \in P$ implies that at least one of the elements a or b is in P, that is, $a \in P$ or $b \in P$.

Suppose that the ideal P is prime according to the first definition, and suppose that $ab \in P$. Then $rab \in P$ for any $r \in D$, so that $(ab) \subset P$. But $(ab) = (a)(b)$, so according to the first definition at least one of the two ideals (a) and (b) is a subset of P. If $(a) \subset P$ then $a \in P$ and if $(b) \subset P$ then $b \in P$. Hence, $a \in P$ or $b \in P$, P is prime also according to the second definition.

Assume now that the ideal P is prime according to the second definition and suppose that $HK \subset P$ where H and K are ideals in P. If at the same time neither H nor K is a subset of P, then H contains an element, to be called a, which is not contained in P, and K contains an element, to be called b, which is not contained in P. But then $ab \in HK$ and so $ab \in P$ although neither a nor b is in P. This is contrary to the assumption that P is prime according to the second definition and shows that either H or K must be a subset of P, that is, P is prime also according to the first definition. We have therefore, proved that the two definitions are equivalent.

It is evident that the unit ideal $D = (1)$ is a prime ideal. The zero ideal $O = (0)$ also is prime for if $ab \in 0$ then $ab = 0$, hence by the definition of an integral domain either $a = 0$ or $b = 0$ (or both), that is, either $a \in O$ or $b \in O$ (or both).

We have seen that as far as the multiplication of ideals is concerned, the ideals $D = (1)$ and $O = (0)$ are analogous to the numbers 1 and 0. Neither one of these numbers is a prime, since 0 was excluded explicitly while 1 is a unit and all units were excepted from the class of primes. Apart from this discrepancy, we shall see in due course that prime ideals are in a very definite way analogous to prime numbers.

2. THE FINITE ASCENDING CHAIN CONDITION

The integral domain D is said to satisfy the *finite ascending chain condition* (to be abbreviated as F.C.C.) if for any sequence of ideals in D, $H_0, H_1, H_2, H_3, \ldots$ for which

$$H_0 \subset H_1 \subset H_2 \subset H_3 \subset \cdots \tag{2.1}$$

there exists a natural number n such that $H_n = H_{n+1} = \ldots$. It amounts to the same to say that in any sequence of ideals of D which satisfies (2.1) the number of different ideals must be finite.

An integral domain D is said to satisfy the *maximum principle* if in every non-empty set, A, of ideals of D there is at least one element, H, which is not contained in any other ideal of A. More formally, for any $K \in A$, $H \subset K$ entails $H = K$. It is not difficult to see that if D does not satisfy the F.C.C. then it cannot satisfy the maximum principle either. For if D does not satisfy the F.C.C. there exists a set of ideals H_n which satisfies (2.1) and such that for every natural number n there exists a natural number $m > n$ for which $H_n \neq H_m$; that is, H_n is *a proper subset* of H_m. But if so then the set $A = \{H_n\}$ is incompatible with the maximum principle, since every one of its elements is a proper subset of another element of A.

Conversely, if D does not satisfy the maximum principle then it does not satisfy the F.C.C. either. For suppose that D does not satisfy the maximum principle. Then there exists a non-empty set A of ideals of D such that for every $H \in A$ there exists a $K \in A$ for which $H \subset K$ but $H \neq K$. We select an element of A calling it H_0. By assumption, there exists another element of A, to be called H_1, such that $H_0 \subset H_1$, but $H_0 \neq H_1$. Similarly, there exists an element of A, to be called H_2, such that $H_1 \neq H_2$ but $H_1 \subset H_2$. Continuing in this way, we obtain an infinite sequence (2.1) such that $H_0 \neq H_1$, $H_1 \neq H_2$, $H_2 \neq H_3$, and so forth. This contradicts the F.C.C.

We have thus shown that the maximum principle and the F.C.C. are equivalent. However, the last argument is less innocuous than it might seem since it includes an infinite number of choices (of H_0, of H_1, of H_2, and so on). It thus involves a special case of the *axiom of choice,* whose status has been the subject of much discussion.

An ideal H in D is called *maximal* if (i) $H \neq D$, and (ii) any ideal K such that $H \subset K$ is equal either to H or to D. In other words, the condition states that there does not exist an ideal which properly contains H and which is properly contained in D.

Theorem. If the ideal H is maximal then it is prime.

Proof. Since H is maximal it must be different from D. It follows that $1 \notin H$. (For if $1 \in H$ then $a = a \cdot 1 \in H$ for all $a \in D$). Let a and b be elements of D such that $ab \in H$ but $b \notin H$. In order to show that H is prime we only have to prove that $a \in H$.

Consider the set

$$K = \{c | c = h + rb,\ h \in H,\ r \in D\} \cdot$$

Thus, K consists of all elements of D that can be written as $h + rb$ where h is any element of H and r is any element of D. Then K is an ideal. (Why?) K contains H as we see by writing $h = h + 0b$ for any $h \in H$. Moreover, K contains b since $b = 0 + 1 \cdot b$. Thus $H \neq K$, K contains H properly. But H is maximal by assumption and so K must be equal to D. Thus, K contains 1. That is to say, there exist $h \in H$ and $r \in D$ such that $1 = h + rb$. Multiplying this equation by a, we obtain

$$a = ah + rab\,.$$

But $ab \in H$, by assumption, so $rab \in H$. And $h \in H$, by construction, so $ah \in H$. Hence, also $a = ah + rab$ is included in H. This proves the theorem.

If every ideal in an integral domain D is principal then D satisfies the F.C.C. For let H_n be a sequence of ideals such that (2.1) is satisfied. Let H be the union of H_n. That is to say, H consists of all elements of D which are contained in at least one H_n,

$$H = \{a | \text{There exists a natural number } n \text{ such that } a \in H_n\}\,.$$

Then H is an ideal. (Why?) By assumption, H contains a particular element a such that $H = (a)$, that is, such that all elements of H are of the form ra (where $r \in D$). But, by the construction of H, a must be contained in H_n for some particular natural number n. And since $H_n \subset H_{n+1} \subset H_{n+2} \subset \ldots$, it follows that a is contained also in $H_{n+1}, H_{n+2}, \ldots$. Hence, for $j = 0, 1, 2, \ldots$, H_{n+j} contains all elements of the form ra, that is, H_{n+j} contains H, $H \subset H_{n+j}$. But since at the same time $H_{n+j} \subset H$ we conclude that $H_{n+j} = H$ and hence, $H_n = H_{n+1} = H_{n+2} = \ldots$. This proves that D satisfies the F.C.C.

In particular, therefore, the ring of rational integers, J, satisfies the F.C.C.

We shall now prove that the ring $J(\sqrt{d})$, where d is square-free and $\sqrt{d}$ is irrational as before, also satisfies the F.C.C.

Suppose first that $d \equiv 2(4)$ or $d \equiv 3(4)$ and let H be an ideal in $J(\sqrt{d})$. Thus, H is a set of numbers $a + b\sqrt{d}$ where a and b are rational integers. Let H_0 be the set of all rational integers a for which there exists a rational integer b such that $a + b\sqrt{d} \in H$. Then H_0 is an ideal in J. Indeed, if $a_1 \in H_0$ and $a_2 \in H_0$, then, for certain rational integers b_1 and b_2, $a_1 + b_1\sqrt{d} \in H_0$ and $a_2 + b_2\sqrt{d} \in H_0$. It follows that $(a_1 + b_1\sqrt{d}) + (a_2 + b_2\sqrt{d})$ is in H, that is $(a_1 + a_2) + (b_1 + b_2)\sqrt{d} \in H$, and similarly, $(a_1 - a_2) + (b_1 - b_2)\sqrt{d} \in H$. Hence, $a_1 + a_2 \in H_0$ and $a_1 - a_2 \in H_0$. At the same time, if r is any rational integer, then it is an element of $J(\sqrt{d})$ and so $r(a_1 + b_1\sqrt{d}) \in H$, that is, $ra_1 + rb_1\sqrt{d} \in H$ and so $ra_1 \in H_0$. This proves that H_0 is an ideal in J. And since all ideals in J are principal there exists a rational integer $a_0 \in H_0$ such that

$$H_0 = \{ra_0 | r \text{ any rational integer}\} .$$

By the definition of H_0, we can find a rational integer b_0 such that $a_0 + b_0\sqrt{d} \in H$.

Now let K be the subset of $J(\sqrt{d})$ which is defined by

$$K = \{k | \text{There exist } h \in H,\ b \in J, \text{ and } r \in J \text{ such that } h = ra_0 + b\sqrt{d} \text{ and } k = h - r(a_0 + b_0\sqrt{d})\} .$$

Since $k = h - r(a_0 + b_0\sqrt{d}) = ra_0 + b\sqrt{d} - r(a_0 + b_0\sqrt{d}) = (b - rb_0)\sqrt{d}$, it follows that all elements of K are of the form $s\sqrt{d}$ where s is a rational integer. K contains the number 0 as we see by putting $h = a_0 + b_0\sqrt{d}$, $b = b_0$ and $r = 1$, so that $k = (b_0 - b_0)\sqrt{d} = 0$. Thus K is not empty. Moreover $K \subset H$ since h and $a_0 + b_0\sqrt{d}$ are both contained in H. Also, if h is any element of H which can be written as $s\sqrt{d}$ for a rational integer s, then $h \in K$. To see this, we only have to put $b = s$ and $r = 0$ in the definition of K, so that $k = s\sqrt{d} = h$. Let

$$H_1 = \{s | s \text{ is a rational integer such that } s\sqrt{d} \in K\} \cdot$$

Then H_1 is again an ideal in J, for if $s_1\sqrt{d} \in K$ and $s_2\sqrt{d} \in K$ then $(s_1 + s_2)\sqrt{d} \in K$ and $(s_1 - s_2)\sqrt{d} \in K$ and if, moreover, r is a rational integer, then $rs_1\sqrt{d} \in K$. We conclude that there exists a rational integer $a_1 \in H_1$ such that

$$H_1 = \{ra_1 | r \text{ any rational integer}\} .$$

Let $b_1 = a_1\sqrt{d}$, then $b_1 \in K$, by the definition of H_1, and so $b_1 \in H$. Let $h_0 = a_0 + b_0\sqrt{d}$ so that h_0 also belongs to H. We are going to show that *b_1 and h_0 together constitute a basis for H in $J(\sqrt{d})$*, that is, $H = (b_1, h_0)$.

Let h be any element of H, $h = a + b\sqrt{d}$. Then there exists a rational integer r such that $a = ra_0$. Moreover, $k = h - rh_0 = (b - rb_0)\sqrt{d}$ belongs to K. Hence $b - rb_0 \in H_1$ and so $b - rb_0 = sa_1$

where s is a rational integer. It follows that $(b - rb_0)\sqrt{d} = sa_1\sqrt{d} = sb_1$, that is, $k = h - rh_0 = sb_1$, $h = sb_1 + rh_0$. Since s and r are integers in $J(\sqrt{d})$, this confirms that h is an element of (b_1, h_0) and proves that $H = (b_1, h_0)$. However, it is quite possible that b_1 or h_0 alone constitutes a basis for H.

Suppose now that $d \equiv 1(4)$. Then $J(\sqrt{d})$ consists of all numbers $\frac{1}{2}(a + b\sqrt{d})$ where a and b are rational integers which are either both odd or both even. Let H be an ideal in $J(\sqrt{d})$ and let H_0 be the set of rational integers a for which there exists a rational integer b such that $\frac{1}{2}(a + b\sqrt{d}) \in H$. Then H_0 is an ideal in J. (Why?) It follows that there exists a rational integer a_0 such that H_0 consists of all numbers ra_0 where ra_0 is any rational integer. By the definition of H_0 there exists a rational integer b_0 such that $h_0 = \frac{1}{2}(a_0 + b_0\sqrt{d}) \in H$.

Let K be the subset of $J(\sqrt{d})$ which is defined by

$$K = \{k | \text{There exist } h \in H,\ b \in J \text{ and } r \in J \text{ such that } h = \tfrac{1}{2}(ra_0 + b\sqrt{d}) \text{ and } k = h - rh_0\}\ .$$

(Compare the corresponding definition for $d \equiv 2(4)$ and $d \equiv 3(4)$.) Then $K \subset H$ since h and h_0 are both contained in H. Also, $k = \frac{1}{2}(ra_0 + b\sqrt{d}) - \frac{1}{2}(ra_0 + rb_0\sqrt{d}) = \frac{1}{2}(b - rb_0)\sqrt{d}$ so that all elements of K are of the form $\frac{1}{2}s\sqrt{d}$ where s is a rational integer. But $\frac{1}{2}s\sqrt{d} = \frac{1}{2}(0 + s\sqrt{d})$ and since 0 is even, this is an integer only if s is even. Thus, $\frac{1}{2}s = t$ is a rational integer, all elements of K are of the form $t\sqrt{d}$ where t is a rational integer. Conversely, any element of H which can be written in the form $t\sqrt{d}$, where t is a rational integer, belongs to K. (Why?) Let H_1 be defined as the set of rational integers t such that $t\sqrt{d} \in K$. Then there exists a rational integer a_1 such that H_1 consists of all numbers of the form ra_1, where r is any rational integer. Let $b_1 = a_1\sqrt{d}$, then $b_1 \in K$ and so $b_1 \in H$. We claim that b_1 and h_0 together constitute a basis for H in $J(\sqrt{d})$.

Let h be any element of H, then we may write $h = \frac{1}{2}(a + b\sqrt{d})$ where a and b are rational integers. There exists a rational integer r such that $a = ra_0$. Let $k = h - rh_0 = \frac{1}{2}(b - rb_0)\sqrt{d}$, then $k \in K$. Hence, $\frac{1}{2}(b - rb_0)$ is a rational integer which belongs to H_1, $\frac{1}{2}(b - rb_0) = sa_1$ where s is a rational integer. It follows that $\frac{1}{2}(b - rb_0)\sqrt{d} = sb_1$, that is, $k = h - rb_0 = sb_1$, $h = sb_1 + rh_0$, $h \in (b_1, h_0)$ and hence $H = (b_1, h_0)$. This proves our assertion also for $d \equiv 1(4)$.

Now let $\{H_n\}$, $n = 0, 1, 2, \ldots$ be a sequence of ideals in the given $J(\sqrt{d})$ such that

$$H_0 \subset H_1 \subset H_2 \subset \cdots$$

as in (2.1). Let H be the union of $\{H_n\}$ so that H is an ideal. (Why?) As we have just shown there exist numbers b_1 and h_0 in H such that $H = (b_1, h_0)$. But then b_1 must be contained in some H_j and all subsequent H_n, and h_0 must be contained in some H_k and all subsequent H_n. It follows that if m is the greater one of the two numbers j and k (or the number $j = k$, if they are equal) then both b_1 and h_0 are contained in H_m and in all H_{m+i} for $i = 1, 2, \ldots$. Thus, for $i = 0, 1, 2, \ldots$, H_{m+i} contains all numbers $sb_1 + rh_0$ where s and r are integers in $J(\sqrt{d})$. This shows that $H \subset H_{m+i}$ for $i = 0, 1, 2, \ldots$. Since, at the same time, $H_{m+i} \subset H$, we conclude that $H = H_m = H_{m+1} = H_{m+2} = \ldots$. Accordingly, $J(\sqrt{d})$ satisfies the F.C.C., as asserted.

3. MORE ABOUT PRIME AND MAXIMAL IDEALS

Let $K = (a)$ be an ideal in J such that $K \neq J$ and $K \neq O$. *Then K is a prime ideal if and only if a is a prime number.*

To prove this assertion, suppose first that a is a prime number in J, and let $b_1b_2 \in K$. That is to say, $b_1b_2 = ra$ for some $r \in J$, a is a divisor of b_1b_2. But if so, then a must be a divisor of either b_1 or b_2 or both; that is, at least one of the two numbers b_1 and b_2 belongs to K. This shows that K is prime.

Suppose now that a is not prime. At any rate a is neither 0 (for then $K = O$) nor a unit (for then $K = J$). Thus, a has a prime divisor, p, such that p is not associated with a, $a = qp$, where q is not a unit. p cannot belong to K for in that case p would be divisible by a hence associated with a (for p is prime). But q cannot belong to K either, for in that case, we should have $q = as$ where $s \in J$. Hence, $a = asp$ and so, by cancellation, $1 = sp$. This would imply that p is a unit, contrary to assumption. Accordingly, the supposition that a is not prime leads to the conclusion that $a = qp \in K$ but $q \notin K$, $p \notin K$; that is, that K is not prime. This completes the proof of our assertion.

We showed earlier that in any integral domain D a maximal ideal is always prime. We shall now show that *in J any prime ideal K other than J or O must be maximal.*

In fact, let K be such an ideal, $K = (p)$, where p is prime as proved above. Suppose that K is not maximal, then there exists an ideal H in J such that $K \subset H$ but $H \neq K$ and $H \neq J$. Let $H = (a)$, then $a \neq 0$ and a is not a unit. At the same time a is a divisor of p since $p \in K$ and $K \subset H$. This shows that a is associated with p. But then $a = \epsilon p$ where ϵ is a unit and so $a \in K$ and $H \subset K$, and hence $H = K$. This is contrary to our assumption on H and proves that K is maximal, as asserted.

We pass to $J(\sqrt{d})$, $d \equiv 2(4)$ or $d \equiv 3(4)$. Let K be an ideal in $J(\sqrt{d})$, and let K' be the set of rational integers which are contained in K. Thus K' is the intersection of K and J, $K' = K \cap J$. K' is an ideal in J. (Why?)

If $K \neq O$, where O is the zero ideal, then K' cannot be equal to the zero ideal either. For let $\beta \in K$ where $\beta \neq 0$. Then β satisfies a quadratic equation

$$x^2 + ax + b = 0$$

where a and b are rational integers. Thus,

$$b = -(\beta^2 + a\beta) = (-\beta - a)\beta$$

which shows that b belongs to K and hence, to K'. If $b \neq 0$ then we have already found a non-zero element of K'. If $b = 0$, then

$$\beta(\beta + a) = 0\,.$$

But $\beta \neq 0$ and so in this case $\beta + a = 0$, $a = -\beta$, $a \in K'$, a is a non-zero element of K'.

If K is prime then K' also is prime.

For let a and b be rational integers such that $ab \in K'$. Then $ab \in K$ and so at least one of the two numbers, shall we say a, belongs to K. But then a belongs to K', as required.

Theorem. If K is a prime ideal in $J(\sqrt{d})$, such that $K \neq J(\sqrt{d})$ and $K \neq O$, then K is maximal.

Proof. With the stated assumptions on K, suppose that K is not maximal. Then there exists an ideal H in $J(\sqrt{d})$ such that $K \subset H$ but $H \neq K$ and $H \neq J(\sqrt{d})$. We define $K' = K \cap J$, as above, and $H' = H \cap J$, so that $H' \supset K'$. (Why?) As we have seen $K' = (p)$ for some prime number p in J.

Now let γ be any number which belongs to H but not to K, so that $\gamma \neq 0$. γ is a root of a quadratic equation

$$x^2 + ax + b = 0\,,$$

where a and b are rational integers. Then $b = (-\gamma - a)\gamma$ so that $b \in H$ and $b \in H'$. If b is not in K, then b is not in K', so that in that case we may conclude that $H' \neq K'$. If b is in K, then, since $b = (-\gamma - a)\gamma$ and K is prime, either γ or $-\gamma - a$ must be in K. But γ is not in K, by assumption, so $-\gamma - a$ and $\gamma + a$ are in K, hence in H, and so $a = (\gamma + a) - \gamma$ is in H and in H'. If a is not in K, then a is not in K', so we may again conclude that $H' \neq K'$. The alternative is that both b and a are in K. But in that case $\gamma^2 = -a\gamma - b$ is in K, and so γ is in K (since K is prime). This contradicts our assumptions on γ, and so the possibility that both a and b are in K must be ruled out. W*e conclude that* $H' \neq K'$.

On the other hand, $H' \neq J$. For $H' = J$ would imply that $1 \in H'$, hence $1 \in H$, hence $H = J(\sqrt{d})$, contrary to assumption. But then H' is an ideal in J which includes K' but which is different from both J and K'. As we have proved already, such an ideal cannot exist since K' is prime in J and is different from 0. Accordingly, the assumption that K is not maximal leads to a contradiction. This proves the theorem.

It will be our main purpose to show that the *ideals* of $J(\sqrt{d})$ satisfy factorization laws rather like the *numbers* of J. These laws are "better" than the factorization laws for the *numbers* of $J(\sqrt{d})$ in which, as we have seen, the factorization into prime numbers need not be unique in the sense detailed previously.

4. IDEALS AND HOMOMORPHISMS

Previously, we introduced ideals in integral domains only. However, the notion can be defined equally well for arbitrary commutative rings. Let R be a commutative ring. Thus R satisfies the first eight rules of Chapter I and also the tenth rule but not necessarily the ninth rule. A non-empty subset K of R is called an *ideal* if for all a_1 and a_2 which belong to K, and for all $r \varepsilon R$, $a_1 + a_2$, $a_1 - a_2$, and ra_1 also belong to K.

Let R_1 and R_2 be two commutative rings. A *mapping* $\phi(x)$ from R_1 into R_2, that is, a function whose argument ranges over R_1 and which takes values in R_2, is called a *homomorphism* if the following conditions are satisfied.

(i) For any a_1 and b_1 in R_1, if $a_2 = \phi(a_1)$ and $b_2 = \phi(a_2)$, then $a_2 + b_2 = \phi(a_1 + b_1)$.

(ii) For any a_1, b_1 in R_1, if $a_2 = \phi(a_1)$ and $b_2 = \phi(a_2)$, then $a_2b_2 = \phi(a_1b_1)$.

Thus, the addition and multiplication of elements of R_1 is mirrored in the addition and multiplication of elements of R_2. An isomorphism between two rings has the same properties but for a homomorphism we do not require that the mapping be one-to-one. Thus, it is now quite possible that several different elements of R_1 are mapped into the same element of R_2. If every element a_2 of R_2 is the image of some element a_1 of R_1, $\phi(a_1) = a_2$, then we say that the homomorphism is from R_1 *onto* R_2 or briefly, that it is *onto*.

We shall denote the zeros of R_1 and R_2 by the same symbol, 0, although strictly speaking it might be more appropriate to employ two distinct symbols. With this notation, we claim that *if* ϕ *is a homomorphism from* R_1 *onto* R_2, *then* $\phi(0) = 0$.

Indeed, if a_1 is any element of R_1, then by (i),

$$\phi(a_1) = \phi(0 + a_1) = \phi(0) + \phi(a_1),$$

that is,

$$\phi(0) + \phi(a_1) = \phi(a_1) .$$

Subtracting $\phi(a_1)$ on both sides of this equation, in other words adding $-\phi(a_1)$, we obtain $\phi(0) = 0$, as asserted.

However, as stated, it is quite possible that there are other elements a_1 of R_1 such that $\phi(a_1) = 0$. Let K_ϕ be the set of these elements for a given homomorphism ϕ from R_1 onto R_2,

$$K_\phi = \{a_1 | \phi(a_1) = 0\} .$$

We *claim that* K_ϕ *is an ideal.*

Indeed, if $a_1 \in K_\phi$ and $b_1 \in K_\phi$, this signifies that $\phi(a_1) = \phi(b_1) = 0$. But then $\phi(a_1 + b_1) = \phi(a_1) + \phi(b_1) = 0 + 0 = 0$ and so $a_1 + b_1 \in K_\phi$. At the same time, $\phi(a_1 - b_1) + \phi(b_1) = \phi((a_1 - b_1) + b_1) = \phi(a_1)$, and so $\phi(a_1 - b_1) = 0$, $a_1 - b_1 \in K_\phi$. Finally, let $r_1 \in R_1$, and $a_1 \in K_\phi$ as before. Then

$$\phi(r_1 a_1) = \phi(r_1)\phi(a_1) = \phi(r_1)0 = 0 ,$$

so that $r_1 a_1 \in K_\phi$. This shows that K_ϕ is an ideal. K_ϕ is called the *kernel* of the homomorphism ϕ.

Now let K be an ideal in a ring R_1. Then we are going to show that there exist a ring R_2 and a homomorphism ϕ from R_1 onto R_2 such that K is the kernel of ϕ, $K = K_\phi$.

It will take us a few steps to reach our aim. We first introduce a relation $x \equiv y$ (read "x is congruent to y") between elements of R_1, as follows. For any element of R_1, a and b, $a \equiv b$ is defined to be true if and only if $a - b$ belongs to K, $a - b \in K$.

The relation $x \equiv y$ *is reflexive.*

For let a be any element of R_1. Then $a - a = 0$ and so $a - a \in K$. It follows that $a \equiv a$.

The relation $x \equiv y$ *is symmetrical.*

For let a and b be any elements of R_1 such that $a \equiv b$. Then $a - b \in K$. But $0 \in K$ and so, by one of the conditions satisfied by an ideal, $0 - (a - b) = b - a \in K$. Hence, $b \equiv a$, as required.

The relation $x \equiv y$ *is transitive.*

For let a, b, and c be any elements of R_1 such that $a \equiv b$ and $b \equiv c$. Then $a - b \in K$ and $b - c \in K$. Hence, $a - c = (a - b) + (b - c) \in K$, since the sum of two elements of K is in K.

Accordingly, we see that the relation of congruence, $x \equiv y$, is what we called an equivalence relation. We may now divide R into non-empty subsets A, B, C, . . . such that two elements of R_1 belong to the same subset if and only if they are congruent. (Why? Compare Chapter III, section 3). One of these sets is K itself. For all elements of K are congruent to one another, and every element of R_1 which is congruent to an element of K, and hence to all elements of K, is included in K. Thus K is one of the sets A, B, C, . . . , in question. They are called the congruence classes of R, (or cosets), with respect to K.

Next, we define a ring R_2 whose elements are the congruence classes of R with respect to K. Let A, B be two such congruence classes. Then we define the sum of A and B as follows. We select arbitrary elements a of A and b of B respectively. Let $c = a + b$. Then c, like any other element of R, must be contained in precisely one congruence class, which may be denoted by C. We call C the *sum of A and B*, $A + B = C$. However, in order to make sure that this is a "good" (unambiguous) definition, we have to verify that the outcome is independent of the particular choice of a and b. Suppose then that a' and b' are another pair of elements, $a' \in A$ and $b' \in B$. Then we have to show that $a' + b' = c'$ belongs to the same congruence class as $a + b = c$, (that is, C). In other words, we have to show that $c \equiv c'$, that is, $c - c' \in K$. But $a \equiv a'$ since a and a' belong to the same congruence class, and similarly $b \equiv b'$. Thus, $a - a' \in K$ and $b - b' \in K$, and hence, $(a - a') + (b - b') \in K$. But $(a - a') + (b - b') = (a + b) - (a' + b') = c - c'$, and so $c - c' \in K$, as required.

Next, we define the product of two congruence classes, A and B. We choose again $a \in A$ and $b \in B$, put $c = ab$, and define the product of A and B, $AB = C$, as the congruence class of c. In order to show that this definition is unambiguous, let $a' \in A$, $b' \in B$, $c' = a'b'$, then we have to show that $c \equiv c'$, that is, that $c - c' \in K$. Now $c - c' = ab - a'b' = ab - ab' + ab' - a'b' = a(b - b') + b'(a - a')$. But $b'(a - a') \in K$ and $a(b - b') \in K$ since $a - a' \in K$ and $b - b' \in K$. Hence the sum $a(b - b') + b'(a - a')$ also belongs to K and this is precisely $c - c'$. This shows that the definition of the product is unambiguous.

We claim that, with these definitions for sum and product, R_2 is a commutative ring. To verify just two of the rules, consider first the associative law of addition. Given any element A, B, and C of R_2, we have to show that $A + (B + C) = (A + B) + C$. Now let a, b, and c be elements of A, B, and C respectively. Then $B + C$ is the congruence class which includes $b + c$ and so $A + (B + C)$ is the congruence class which includes $a + (b + c)$. Similarly, $(A + B) + C$ is the congruence class (element of R_2) which includes $(a + b) + c$. But then $A + (B + C)$ and $(A + B) + C$ coincide since $a + (b + c) = (a + b) + c$.

Next, we are going to show that there exists a neutral element with respect to addition. Indeed, this is precisely K itself. For let A be any element of R_2. To find $A + K$ we choose an arbitrary $a \epsilon A$ and we choose 0 in K. Then $a + 0 = a$, and so $A + K$ is the congruence class which includes a, that is, A itself. This proves that $A + K = A$.

The remaining rules for a commutative ring also are satisfied by R_2. (Why?)

Next, we shall show that there exists a homomorphism ϕ from R_1 onto R_2. Let a be any element of R_1. Then we define $\phi(a)$ as *the* congruence class A which includes a, $\phi(a) = A$. ϕ is a homomorphism. For let $\phi(a) = A$, $\phi(b) = B$, then $\phi(a + b)$ is the congruence class which includes $a + b$. But, by the definition of the sum in R_2 this is precisely $A + B$, and so $\phi(a + b) = A + B = \phi(a) + \phi(b)$. Similarly $\phi(ab)$ is the congruence class which includes ab and this is AB by the definition of the product in R_2. Hence $\phi(ab) = AB = \phi(a)\phi(b)$. This shows that ϕ is a homomorphism. The homomorphism is *onto* for if $A \epsilon R_2$ then A is a non-empty class of R_1. Let a be any element of A, then $\phi(a) = A$ so that A is included in the range of 0. This shows that ϕ is onto.

Finally, let us find the kernel of ϕ, K_ϕ. By definition, K_ϕ consists of all elements of R_1 which are mapped by ϕ on the zero of R_2. But the zero of R_2 is K, so K_ϕ consists of all elements k of R_1 such that $\phi(k) = K$. But $\phi(k) = K$ precisely when $k \epsilon K$. We conclude that $K_\phi = K$. Thus, we have constructed a homomorphism ϕ from R_1 onto another ring such that the kernel of ϕ coincides with the given ideal K, as promised.

We have used the notation $x \equiv y$ in order to denote the relation introduced above for a given ideal K. It is customary to emphasize the fact that the relation depends on K by writing more explicitly $x \equiv y(K)$ or $x \equiv y \bmod K$ (read "x is congruent to y modulo K"). This is rather like the notation and terminology used previously in connection with the definition of residue rings as in Chapter III, section 3. Indeed, the residue rings introduced there are easily shown to coincide with the rings R_2, introduced above, if R_1 is the ring of rational integers J, and K is an ideal in J, $K = (n)$ where n is a positive natural number. Then the relation of congruence used in the present section coincides with the relation $x \equiv y(n)$ of Chapter III, section 4, since $a - b \epsilon (n)$ if and only if $a - b$ is divisible by n. The congruence classes for the given K are precisely the classes $A_0, A_1, \ldots, A_{n-1}$ of Chapter III, and the operations of addition and multiplication introduced there also are special cases of the definitions given above.

The connection between ideals and homomorphism will not be made use of in the further developments given in the present book. It is, however, a basic aspect of the theory of ideals and has many applications in other branches of mathematics.

EXERCISES

1. Let K_1, K_2, K_3 be three ideals in J such that $K_1 \subset K_2 \subset K_3$, $K_1 \neq K_2$, $K_2 \neq K_3$, and $K_1 = (6)$. Prove that $K_3 = J$.

2. Find all ideals in the ring S defined by case **i.**, exercise 2, Chapter 3, page 39.

3. Find all ideals in the ring of congruence classes modulo 10.

4. Prove that a ring which contains just two different ideals is a field.

VII
Polynomials

1. RINGS OF POLYNOMIALS

You have come across polynomials before in one form or another. In the present book polynomials have appeared on the left-hand side of *polynomial equations*, more particularly, of quadratic equations

$$x^2 + ax + b = 0.$$

In this section we shall concentrate our attention on these polynomials as such. The arithmetic of polynomials has many applications which are outside our context.

Let D be any commutative ring with unit element. In particular, D may be an integral domain, for example J or some $J(\sqrt{d})$. By a polynomial in D we mean any symbolic expression

(1.1)
$$p(x) = a_0 + a_1x + a_2x^2 + \cdots + a_nx^n,\ n \geq 0,\ a_i \epsilon D \text{ for } 0 \leq i \leq n$$

where $a_n \neq 0$. However, when $n = 0$, we permit $a_0 = 0$, so that 0 alone also counts as a polynomial. The *degree* of the polynomial n is called and $a_0, a_1, \ldots, a_n$ are its coefficients. The polynomials $p(x) = a_0 + a_1x + \ldots + a_nx^n$ and $q(x) = b_0 + b_1x + \ldots + b_mx^m$ are *equal* if $n = m$ and $a_i = b_i$ for $0 \leq i \leq n$.

When writing down a polynomial it is permissible, by universal agreement, to omit terms $0x^m$ and to add such terms, where convenient. Thus $2x^3$ is written out in full, the polynomial $0 + 0x + 0x^2 + 2x^3$, and $1 + 2x + 0x^2$ is $1 + 2x$. Similarly, in place of $1x^n$ we may write also x^n.

As (1.1) stands, it is a purely formal expression. However, in determining addition and multiplication for polynomials we treat $p(x)$ *as if it were* the sum of a_0 and $a_1x + a_2x^2$ etc., and we treat each term (or monomial) $a_m x^m$ *as if it were* the product of a_m and of the nth power of x. We also assume that the polynomials satisfy the associative, commutative, and distributive laws, just like the ring D itself. With these ideas as *guidelines*, we define the sum of two polynomials, $p(x)$, as given by (1.1) and

$$q(x) = b_0 + b_1 x + b_2 x^2 + \cdots + b_m x^m \tag{1.2}$$

for $m = n$ by

$$p(x) + q(x) = (a_0 + b_0) + (a_1 + b_1)x + (a_2 + b_1)x^2 + \cdots + (a_n + b_n)x^n . \tag{1.3}$$

If the degrees of the two polynomials are unequal, then we add terms of the form $0x^j$ to one of them to make its degree apparently equal to that of the other polynomial. Equivalently, we simply omit the corresponding a_j or b_j in (1.3). For example,

$$(3 + 2x + (-\tfrac{3}{2})x^2) + (\tfrac{1}{2} + 5x + 2x^2 + 5x^3) = (3 + 2x + (-\tfrac{3}{2})x^2 + 0x^3) + (\tfrac{1}{2} + 5x + 2x^2 + 5x^3) = \tfrac{7}{2} + 7x + \tfrac{1}{2}x^2 + 5x^3 . \tag{1.4}$$

The product of two polynomials $p(x)$ and $q(x)$ is obtained formally by multiplying out and rearranging in rising powers of x, so

$$p(x)q(x) = (a_0 + a_1 x + a_2 x^2 + \cdots + a_n x^n)(b_0 + b_1 x + b_2 x^2 + \cdots + b_m x^m) = a_0 b_0 + (a_0 b_1 + a_1 b_0) + (a_0 b_2 + a_1 b_1 + a_2 b_0)x^2 + \cdots + a_n b_m x^{n+m} = c_0 + c_1 x + c_2 x^2 + \cdots + c_{n+m} x^{n+m}. \tag{1.5}$$

In this formula, the coefficient c_j of the product is given by

$$c_j = a_0 b_j + a_1 b_{j-1} + a_2 b_{j-2} + \cdots + a_j b_0 \quad 0 \leq j \leq n + m , \tag{1.6}$$

where a_i, b_i on the right-hand side have to be set equal to 0 for $i > n$ and $i > m$, respectively.

With these definitions of sum and product the set of polynomials under consideration becomes a commutative ring with unit element. (Why? Check to see that the first ten rules of Chapter 1 are satisfied.) This ring is denoted by $D[x]$. The zero and unit elements of $D[x]$ are 0 and 1 regarded as polynomials. The inverse of the polynomial $p(x)$ (see (1.1)) with respect to addition is

$$-p(x) = (-a_0) + (-a_1)x + (-a_2)x^2 + \cdots + (-a_n)x^n \cdot \tag{1.7}$$

The ring $D[x]$ contains the ring of *constant* polynomials, whose degree, n, is equal to 0. This ring is isomorphic to D. Or, if we admit a certain latitude of expression, we may identify a constant *polynomial*, a_0, with the *element* of D, a_0, so that in this sense $D[x]$ contains the ring D. In this sense also, the notation (1.1) is justified in retrospect, since $p(x)$ is indeed the sum of a_0 and of the product of a_1 multiplied by the polynomial $x = 0 + 1 \cdot x$, and of a_2, multiplied by x, multiplied by x, and so forth. We may deduce, although with more effort than is apparent from the present sketch, that if a is an element of D or of some larger ring which includes D, and $p(x)$ and $q(x)$ are polynomials in D, as given by (1.1) and (1.2), and $r(x) = p(x) + q(x)$, $s(x) = p(x)q(x)$, then $r(a) = p(a) + q(a)$, $s(a) = p(a)q(a)$, where $p(a)$, $q(a)$, $r(a)$, $s(a)$ are obtained by substituting a in $p(x)$, $q(x)$, $r(x)$, $s(x)$, respectively. Putting it in a different way we may say that we obtain the same result by substituting a in the sum (product) of $p(x)$ and $q(x)$ as we do by adding (multiplying) the elements obtained by substituting a in $p(x)$ and in $q(x)$.

The degree of a polynomial $p(x)$ was defined in connection with (1.1), and we shall denote it by deg $p(x)$. Suppose now that D is an integral domain and that $p(x)$ and $q(x)$ are two polynomials of $D[x]$ neither of which is (*the polynomial*) 0. Then

$$\text{(1.8)} \qquad \deg\,[p(x)q(x)] = \deg p(x) + \deg q(x)$$

and the coefficient of the highest power of x in $p(x)q(x)$ is the product of the coefficient of the highest powers of x in $p(x)$ and $q(x)$ respectively. Indeed, if $p(x)$ and $q(x)$ are given by (1.1) and (1.2), where $a_n \neq 0$, and $b_m \neq 0$, then it is easy to verify that the coefficient of the highest power of $p(x)q(x)$ is $c_{n+m} = a_n b_m \neq 0$. It follows that $p(x)q(x) \neq 0$ and hence, that $D[x]$ also is an integral domain.

Suppose, in particular, that D is a field. As we have just seen $D[x]$ is then an integral domain. Then

The units of $D[x]$ are precisely the constant polynomials other than 0.

For if $p(x) = a$ where $a \in D$ and $a \neq 0$, then $[p(x)]^{-1} = 1/a$. And if $p(x)$ is not a constant, then deg $p(x) > 0$ and (1.8) then shows that $p(x)$ cannot have a reciprocal. (Why?)

Theorem. Let $p(x)$ and $q(x)$ be polynomials of $D[x]$ such that $p(x) \neq 0$. Then there exists polynomials $s(x) \in D[x]$ and $r(x) \in D[x]$ such that

$$\text{(1.9)} \qquad q(x) = s(x)p(x) + r(x)$$

where $r(x)$ is either 0 or deg $r(x) <$ deg $p(x)$.

This is the *remainder theorem*. Formula (1.9) is analogous to division with remainder for natural numbers.

To prove the theorem, we observe that if deg $q(x) <$ deg $p(x)$ then we may satisfy (1.9) by putting $s(x) = 0$, $r(x) = q(x)$. This shows that for given $p(x)$ the conclusion of the theorem can be satisfied for all polynomials $q(x)$ whose degree does not exceed [deg $p(x)$] $+ j$ for $j = 0$. We now suppose that we have proved the conclusion of the theorem for all polynomials $q(x)$ whose degree does not exceed [deg $p(x)$] $+ j$; and we proceed to show that we may establish the same conclusion for all polynomials $q(x)$ whose degree is *exactly* [deg $p(x)$] $+ j + 1$ and hence, for all polynomials $q(x)$ whose degree *does not exceed* [deg $p(x)$] $+ j + 1$. This will constitute a proof of the theorem by induction.

Suppose then that $p(x)$ and $q(x)$ are given by (1.1) and 1.2) so that $m = \deg q(x) = [\deg p(x)] + j + 1 = n + j + 1$. The polynomial

$$p_1(x) = \frac{b_m}{a_n} x^{j+1} p(x)$$

is of degree $n + j + 1 = m$ and the coefficient of x^m in $p_1(x)$ is b_m, that is, the same as the coefficient of x^m in $q(x)$. It follows that the degree of the polynomial $q_1(x) = q(x) - p_1(x)$ is smaller than m and hence that it does not exceed [deg $p(x)$] $+ j$. Thus, by the assumption of our induction there exist polynomials $s_1(x)$ and $r_1(x)$, where $r_1(x)$ is either 0 or of degree less than n, such that

$$q_1(x) = s_1(x)p(x) + r_1(x) .$$

But then

$$q(x) = q_1(x) + p_1(x) = s_1(x)p(x) + r_1(x) + \frac{b_m}{a_n} x^{j+1} p(x)$$

$$= \left(s_1(x) + \frac{b_m}{a_n} x^{j+1}\right) p(x) + r_1(x) .$$

This shows that $q(x)$ can be written in the required form and proves the theorem.

2. ROOTS OF POLYNOMIALS

Suppose that the field D is contained in a larger field F. An element α of F is *algebraic with respect to D*, if there exists a polynomial of positive degree $p(x) \in D[x]$, such that α is a *root* of $p(x)$, that is, such that $p(\alpha) = 0$. For example, if F is the field of complex numbers and D the field of rational numbers, $D = Ra$ in our earlier notation, then the elements of F which are algebraic with respect to D are the *algebraic numbers*. In particular, all elements of $Ra(\sqrt{d})$ are algebraic.

A polynomial $p(x)$ of *positive degree*, deg $p(x) \geq 1$, is called *irreducible* if it is not the product of two polynomials of positive degree. The irreducible polynomials play the part of the prime elements in the arithmetic of polynomials. However, we shall not develop that theory here.

Let α be any algebraic number. *Then there exists an irreducible polynomial $q(x)$ such that $q(\alpha) = 0$.* Moreover, any other irreducible polynomial which has α as a root can be obtained from $q(x)$ by multiplication by a constant (rational number).

For since α is algebraic there exist polynomials with rational coefficients and of positive degree which have α as a root. Among them, let $q(x)$ be one whose degree is as small as possible. The polynomial $q(x)$ is irreducible for if $q(x) = r(x)s(x)$, where both $r(x)$ and $s(x)$ have positive degrees, then $\deg q(x) = \deg r(x) + \deg s(x)$ and hence, $\deg r(x) < \deg q(x)$ and $\deg s(x) < \deg q(x)$. Also

$$r(\alpha)s(\alpha) = q(\alpha) = 0$$

and so either $r(\alpha) = 0$, or $s(\alpha) = 0$, or both. Thus either $r(x)$ or $s(x)$ is a polynomial whose degree is positive, but smaller than the degree of $q(x)$ and which has α as a root. This contradicts our choice of $q(x)$. We conclude that $q(x)$ is irreducible.

Now let $p(x)$ be any other polynomial such that $p(\alpha) = 0$. We apply the remainder theorem, so,

$$p(x) = s(x)q(x) + r(x)$$

where the degree of $r(x)$ is less than the degree of $q(x)$. Then

$$p(\alpha) = s(\alpha)q(\alpha) + r(\alpha)$$

and so

$$r(\alpha) = 0 .$$

This is possible only if $r(x) = 0$. (Why?) Hence, $p(x) = s(x)q(x)$ so that $q(x)$ is a divisor of all polynomials with root α. It follows that $p(x)$ can be irreducible only if $\deg p(x) = \deg q(x)$, in which case $s(x)$ is of degree 0, that is, $s(x)$ is a constant.

Let $p(x)$ be given by (1.1) where $a_0, a_1, \ldots, a_n$ are rational integers, $n \geq 0$, $a_n \neq 0$. $p(x)$ is called *primitive* if the only divisors common to all the coefficients of $p(x)$ are 1 and -1 (that is, the units of J). Thus, $1 + 6x + 12x^2$ is primitive, $3 + 6x + 12x^2$ is not primitive. We are going to prove the following important theorem, which is due to C. F. Gauss (1777–1855).

Theorem. The product of two primitive polynomials is primitive.

Proof. Let the primitive polynomials $p(x)$ and $q(x)$ be given by (1.1) and (1.2) respectively, $n \geq 0$, $m \geq 0$, $a_n \neq 0$, $a_m \neq 0$. Then the product of $p(x)$ and $q(x)$ is of degree $n + m$, so

$$r(x) = p(x)q(x) = c_0 + c_1x + c_2x^2 + \cdots + c_{n+m}x^{n+m}, \; c_{n+m} \neq 0 .$$

Now suppose, contrary to the assertion of the theorem, that $r(x)$ is not primitive. In that case $c_0, c_1, \ldots, c_{n+m}$ have a common divisor, c, which is not a unit. c in turn must be divisible by some *prime number*, d. Then d also is a common divisor of $c_0, c_1, c_2, \ldots, c_{n+m}$. But $p(x)$ is primitive, so not all coefficients of $p(x)$ are divisible by d. It follows that there exists a smallest natural number k such that a_k is not divisible by d. Similarly, there exists a smallest natural number l such that b_l is not divisible by d. Now consider the coefficient c_{k+l} of $r(x)$. In terms of the coefficient of $p(x)$ and $q(x)$, c_{k+l} is given by

$$c_{k+l} = a_0b_{k+l} + a_1b_{k+l-1} + \cdots + a_{k-1}b_{l+1} + a_kb_l + a_{k+1}b_{l-1} + \cdots + a_{k+l}b_0 ,$$

where a_i and b_i are set equal to 0 for $i > n$ and $i > m$ respectively. Hence,

$$a_kb_l = c_{k+l} - a_0b_{k+l} - a_1b_{k+l-1} \cdots - a_{k-1}b_{l+1} - a_{k+1}b_{l-1} \cdots - a_{k+l}b_0 .$$

On the right-hand side of this equation c_{k+} is divisible by d, by assumption; $a_0b_{k+l}, a_1b_{k+l-1}, \ldots, a_{k-1}b_{l+1}$ are divisible by d because their first factor is divisible by d; and $a_{k+1}b_{l-1}, \ldots, a_{k+l}b_0$ are divisible by d because their second factor is divisible by d. It follows that the right-hand side of the equation is divisible by d. But the left-hand side cannot be divisible by d since neither a_k nor b_l is divisible by d and d is prime. This contradiction shows that $c_0, c_1, \ldots, c_{n+m}$ cannot have a common divisor other than 1 and -1 and proves that $r(x)$ is primitive, as asserted.

Now let $p(x)$ and $q(x)$ be polynomials which are given by (1.1) and (1.2) where $n \geq 1$, $m \geq 1$, $a_n = 1$, $b_m \neq 0$, and where the coefficients, a_i, b_i, are rational integers. Let

$$r(x) = c_0 + c_1x + \cdots + c_lx^l ,$$

where $l \geq 0$, $c_l \neq 0$, and where the c_i are rational but not necessarily integers. Suppose that $q(x)r(x) = p(x)$. *Then we propose to show that b_m is a divisor of $b_0, b_1, \ldots, b_{m-1}$.*

By assumption

$$(2.1) \quad (b_0 + b_1x + \cdots + b_mx^m)(c_0 + c_1x + \cdots + c_lx^l) = a_0 + a_1x + \cdots + a_{n-1}x^{n-1} + x^n .$$

We write the c_i as fractions with least common denominators, s, $c_i = t_i/s$. By its definition, s has no factor other than 1 or -1 in common with all t_i. Multiplying (2.1) by s we obtain

$$(2.2)\quad (b_0 + b_1x + \cdots + b_mx^m)(t_0 + t_1x + \cdots + t_lx^l) = sa_0 + sa_1x + \cdots + sa_{n-1}x^{n-1} + sx^n .$$

Let b be the greatest natural number which divides $b_0, b_1, \ldots, b_m$, simultaneously, $b \geq 1$ (b is the *greatest common divisor* of $b_0, b_1, \ldots, b_m$). Since $s = b_mt_l$, b is also a divisor of s. We write $b_i = b_i'b$, $i = 0, 1, \ldots, m$, and $s = s'b$. Dividing (2.2) by b, we then obtain

$$(2.3)\quad (b_n' + b_1'x + \cdots + b_m'x^m)(t_0 + t_1x + \cdots + t_lx^l) = s'a_0 + s'a_1x + \cdots + s'x^n .$$

$b_0', b_1', \ldots, b_m'$ cannot have any common divisor other than 1 and -1 for if $b' > 1$ were a common divisor of these numbers then $b'b$ would be a common divisor of $b_0, b_1, \ldots, b_m$, and this is impossible since $b'b$ is greater than b. Thus, $b_0' + b_1'x + \cdots + b_m'x^m$ is primitive. But $t_0 + t_1x + \cdots + t_lx^l$ must be primitive as well, for every common divisor t of $t_0, t_1, \ldots, t_l$ must be a divisor of $s' = b_m't$ and hence of $s = b_m't_lb$. But then $t = \pm 1$ otherwise s would not be the least common denominator of $c_0, c_1, \ldots, c_l$. It follows that $t_0 + t_1x \cdots + t_lx^l$ is primitive and hence, by the last theorem, that $s'a_0 + s'a_1x + \cdots + s'x^m$ is primitive. But this is possible only if $s' = \pm 1$, and so $s = s'b = \pm b$. Moreover, $b_m't_l = s' = \pm 1$ shows that $b_m' = \pm 1$ (since b_m' is a unit in J). Hence $b_m = b$ or $b_m = -b$. But b is a divisor of $b_0, b_1, \ldots, b_{m-1}$ and so we have shown that b_m is indeed a divisor of $b_0, b_1, \ldots, b_{m-1}$, as asserted.

From this result, we draw the following conclusion.

Let β be an element of a quadratic field $Ra(\sqrt{d})$ such that β is a root of a polynomial with rational integer coefficients

$$(2.4)\qquad p(x) = a_0 + a_1x + \cdots + a_{n-1}x^{n-1} + x^n,\ n \geq 1 ,$$

in which the coefficient of the highest power of x is 1. (Such a polynomial is called *monic.*) *Then β is an algebraic integer.*

The point of this assertion is that an algebraic integer in a quadratic field was defined previously as a number which is a root of a polynomial such as (2.4) *for $n = 2$.*

Since β is an element of a quadratic field, we know that it is a root of some irreducible polynomial $q(x) = b_0 + b_1x + b_2x^2$ whose coefficients are rational integers, not all 0. Hence, as proved already, there exists a polynomial $r(x)$ which belongs to $Ra[x]$, that is, whose coefficients are rational, such that $p(x) = q(x)r(x)$. If $b_2 \neq 0$ then our previous conclusion shows that $b_2|b_1$, and $b_2|b_0$, that is, $b_0 = b_0'b_2$, $b_1 = b_1'b_2$, $q(x) = b_2(b_0' + b_1'x + x^2)$ where b_0', b_1' are rational integers. Thus, $b_0' + b_1'\beta + \beta^2 = 0$ so that β is an algebraic integer according to our definition. If

$b_2 = 0$ we must have $b_1 \neq 0$, and in that case $b_1 | b_0$, $b_0 = b_0' b_1$, $q(x) = b_1(b_0' + x)$ where b_0' is a rational integer. Then $b_0' + \beta = 0$, so that β is even a rational integer. Thus, β is an algebraic integer whenever it is a root of a polynomial such as (2.4).

So far, we have assumed that the coefficients $a_0, a_1, \ldots, a_{n-1}$ in (2.4) are rational integers. We shall now suppose only that they are (algebraic) integers in the given $Ra(\sqrt{d})$ of which β is an element. *Then if* $p(\beta) = 0$, β *is an algebraic integer.*

To prove our assertion recall that in Chapter IV, section 5, we introduced for any element $\gamma = c + b\sqrt{d}$ of $Ra(\sqrt{d})$ where c and b are rational, its *conjugate*, denoted by $\bar{\gamma}$, which was defined by $\bar{\gamma} = c - b\sqrt{d}$. Then $\bar{\bar{\gamma}} = c + b\sqrt{d} = \gamma$, and if $\gamma_1 = c_1 + b_1\sqrt{d}$, $\gamma_2 = c_2 + b_2\sqrt{d}$, then

$$\begin{aligned} \overline{\gamma_1 + \gamma_2} &= \overline{(c_1 + b_1\sqrt{d}) + (c_2 + b_2\sqrt{d})} = \overline{(c_1 + c_2) + (b_1 + b_2)\sqrt{d}} \\ &= (c_1 + c_2) - (b_1 + b_2)\sqrt{d} = (c_1 - b_1\sqrt{d}) + (c_2 - b_2\sqrt{d}) = \bar{\gamma}_1 + \bar{\gamma}_2 \end{aligned}$$

and similarly, $\overline{\gamma_1\gamma_2} = \bar{\gamma}_1\bar{\gamma}_2$.

If γ is rational, then $b = 0$ and so $\gamma = \bar{\gamma}$. Conversely, if $\gamma = \bar{\gamma}$ then $c + b\sqrt{d} = c - b\sqrt{d}$, so $b = 0$ and $\gamma = c$ is rational. Thus, for γ to be rational it is necessary and sufficient that $\gamma = \bar{\gamma}$.

Define the polynomial $\bar{p}(x)$ by

$$\bar{p}(x) = \bar{a}_0 + \bar{a}_1 x + \cdots + \bar{a}_{n-1}x^{n-1} + x^n \tag{2.5}$$

and consider the product $q(x) = p(x)\bar{p}(x)$. Evidently, $q(\beta) = p(\beta)\bar{p}(\beta) = 0$. We shall show that the coefficients of $q(x)$ are rational integers.

We write

$$q(x) = c_0 + c_1 x + c_2 x^2 + \cdots + c_{2n}x^{2n},$$

where

(2.5)

$$c_k = a_0\bar{a}_k + a_1\bar{a}_{k-1} + a_2\bar{a}_{k-2} + \cdots + a_{k-1}\bar{a}_1 + a_k\bar{a}_0, \quad 0 \leq k \leq n.$$

Whenever an a_i appears in this formula with subscript i greater than n, we put $a_i = 0$ (so that $\bar{a}_i = 0$) while for $i = n$, $a_n = \bar{a}_n = 1$.

We see immediately that the c_k are integers, for they are sums of products of integers. It remains to be shown that c_k is rational. For this purpose, we only have to verify that $\bar{c}_k = c_k$. Now taking the conjugate on the right-hand side of (2.5) and using the fact that the conjugate of a sum (product) of certain numbers is the sum (product) of the conjugates of these numbers (see above), we obtain

$$\begin{aligned}\bar{c}_k &= \overline{a_0\bar{a}_k + a_1\bar{a}_{k-1} + \cdots + a_{k-1}\bar{a}_1 + a_k\bar{a}_0} \\ &= \overline{a_0\bar{a}_k} + \overline{a_1\bar{a}_{k-1}} + \cdots + \overline{a_{k-1}\bar{a}_1} + \overline{a_k\bar{a}_0} \\ &= \bar{a}_0\bar{\bar{a}}_k + \bar{a}_1\bar{\bar{a}}_{k-1} + \cdots + \bar{a}_{k-1}\bar{\bar{a}}_1 + \bar{a}_k\bar{\bar{a}}_0 \\ &= \bar{a}_0 a_k + \bar{a}_1 a_{k-1} + \cdots + \bar{a}_{k-1}a_1 + \bar{a}_k a_0 \,.\end{aligned}$$

But the last expression for $\bar{c}_k$ is exactly the same as the right-hand side of (2.5) when written in the reverse order. It follows that $\bar{c}_k = c_k$, c_k is rational and hence, is a rational integer. Moreover, the coefficient of the highest power of x which appears in $q(x)$ is $c_{2n} = a_n\bar{a}_n = 1$. Since $q(\beta) = 0$ we may therefore conclude from our previous result that β is an integer.

3. ANOTHER TEST FOR ALGEBRAIC INTEGERS

Let β be an element of the quadratic field $Ra(\sqrt{d})$ which satisfies the following condition. There exist a number γ in $Ra(\sqrt{d})$ and a sequence of numbers $\mu_1, \mu_2, \mu_3, \ldots$, which belong to $J(\sqrt{d})$, that is, which are integers in $Ra(\sqrt{d})$, such that $\beta^j = \mu_j\gamma$, $j = 1, 2, 3, \ldots$. *Then β is an integer.*

Observe that γ is not assumed to be an integer, otherwise the conclusion would be trivial.

To prove the assertion, consider the sequence of ideals $\{K_m\}$ in $J(\sqrt{d})$, $m = 1, 2, \ldots$, in which $K_m = (\mu_1, \mu_2, \ldots, \mu_m)$. Thus K_m consists of all numbers which can be written in the form $\alpha_1\mu_1 + \alpha_2\mu_2 + \cdots + \alpha_m\mu_m$, where $\alpha_1, \alpha_2, \ldots, \alpha_m$ are integers. Then

$$K_1 \subset K_2 \subset K_3 \subset \cdots. \quad \text{(Why?)}$$

Now we know that $J(\sqrt{d})$ satisfies the finite ascending chain condition, so there exists a natural number m such that $K_m = K_{m+1} = \cdots$. It then follows that μ_{m+1}, which is contained in K_{m+1}, is contained in K_m. Hence, there exist numbers $\alpha_1, \ldots, \alpha_m$ which belong to $J(\sqrt{d})$ such that

$$\mu_{m+1} = \alpha_1\mu_1 + \alpha_2\mu_2 + \cdots + \alpha_m\mu_m \,.$$

Multiplying this equation by γ we obtain

$$\mu_{m+1}\gamma = \alpha_1\mu_1\gamma + \alpha_2\mu_2\gamma + \cdots + \alpha_m\mu_m\gamma$$

or, which is the same,

$$\beta^{m+1} = \alpha_1\beta + \alpha_2\beta^2 + \cdots + \alpha_m\beta^m \,.$$

We conclude that β is a root of the monic polynomial,

$$p(x) = -\alpha_1 x - \alpha_2 x^2 - \cdots - \alpha_m x^m + x^{m+1} \,,$$

whose coefficients belong to $J(\sqrt{d})$. The last result of section 2 above now shows that β is an integer, as asserted.

The same conclusion is true if β if an element of *Ra* (that is, a rational number) and there exists a *rational* γ such that $\beta^j = \mu_j\gamma$, $j = 1, 2, 3, \ldots$, where the μ_j are all *rational* integers. In fact this may be regarded simply as a special case of the result just proved. (Why?) Alternatively, it may be established directly.

EXERCISES

1. Find the G.C.D. of the two polynomials $x^6 - 1$ and $x^4 + 2x^3 + 3x^2 + 2x + 1$.

2. Show that the polynomial $1 + x^2$ is prime in the ring of polynomials with real coefficients.

3. D is the ring of polynomials with real coefficients and is mapped on a ring F by a homomorphism with kernel $(1 + x^2)$. Show that F is (or, is isomorphic to) the field of complex numbers.

4. D is the ring of polynomials with rational coefficients and is mapped on a ring F by a homomorphism with kernel $d + x^2$ where d is irrational. Show that F is isomorphic to the field $Ra\ (\sqrt{d})$.

VIII
Factorization of Ideals

1. PREPARATION

In this chapter, which is our last, we shall show that the ideals of J and of $J(\sqrt{d})$ obey factorization laws which are rather like the factorization laws for natural numbers. More precisely, we shall see that the ideals satisfy a unique factorization law for *all* quadratic fields. Thus, the situation for ideals is "better" than for the integers themselves in some quadratic fields since we have shown that there exist fields in which the factorization into primes is definitely not unique (even if we identify associated numbers).

All the general results given in this chapter are true equally for J and for $J(\sqrt{d})$. In order to avoid the need for perpetual repetitions we shall agree from now on that the symbol D shall either denote J, in which case the symbol F shall denote the field of rational numbers Ra, or D shall denote some $J(\sqrt{d})$, in which case F stands for $Ra(\sqrt{d})$. You will verify without difficulty that our arguments hold under all these interpretations of D and F.

Theorem. For every ideal K in D where $K \neq O$, there exists a set of prime ideals $P_1, P_2, \ldots, P_m$ in D, $m \geq 1$, such that $K \subset P_i$, $i = 1, 2, \ldots, m$ and

$$P_1 P_2 \cdots P_m \subset K,$$

that is, the product of the prime ideals P_i is contained in K.

Proof. For a given D, let A be the set of ideals K in D for which there does not exist any set of prime ideals, $P_1, \ldots, P_m$ which satisfies the conclusion of the theorem. If A is empty then we have finished, for in that case all ideals of D have the required property. If A is not empty then, by the maximum principle, which is satisfied by D, there exists an ideal K in A which is not contained in other ideals of K. K cannot be prime, for if it were, then $P_1 = K$ would satisfy the conclusions of the theorem. Thus there exist a, b, $a \epsilon D$, $b \epsilon D$, such that $a \notin K$, $b \notin K$ but $ab \epsilon K$. Define the subsets H_1 and H_2 of D by

$$H_1 = \{h_1 | h_1 = k + ra \text{ for some } k \epsilon K,\ r \epsilon D\}$$
$$H_2 = \{h_2 | h_2 = k + rb \text{ for some } k \epsilon K,\ r \epsilon D\}\ .$$

Thus, H_1 consists of all numbers of D that can be written in the form $h_1 = k + ra$ and H_2 consists of all numbers $h_2 = k + rb$ where $k \epsilon K$ and $r \epsilon D$.

Both H_1 and H_2 are ideals in D as we saw previously in similar cases. (Why?) Now let $h_1 \epsilon H_1$ and $h_2 \epsilon H_2$ so that $h_1 = k_1 + r_1a$, $h_2 = k_2 + r_2b$ where $k_1 \notin K$, $k_2 \notin K$. Then

$$h_1h_2 = (k_1 + r_1a)(k_2 + r_2b) = k_1k_2 + r_1ak_2 + r_2bk_1 + r_1r_2ab\ ,$$

and this belongs to K since ab belongs to K. Since H_1H_2 consists of sums of products of this kind we conclude that $H_1H_2 \subset K$. On the other hand, $K \subset H_1$, and $K \subset H_2$ (Why?) but $H_1 \neq K$ since a is contained in H_1 but not in K_1, and $H_2 \neq K$ since b is contained in H_2 but not in K. It follows that neither H_1 nor H_2 are contained in A. Thus, there exist prime ideals $R_1, R_2, \ldots, R_j$, and $S_1, S_2, \ldots, S_l$, $j \geq 1$, $l \geq 1$, such that $H_1 \subset R_i$, $i = 1, \ldots, j$, and $H_2 \subset S_i$, $i = 1, \ldots, l$, and $R_1R_2 \cdots R_j \subset H_1$, and $S_1S_2 \cdots S_l \subset H_2$. But then

$$R_1R_2 \cdots R_jS_1S_2 \cdots S_l \subset H_1H_2 \subset K \text{ and } K \subset R_i,$$
$$i = 1, \ldots, j, \text{ and } K \subset S_i,\ i = 1, \ldots, l.$$

This shows that the conclusion of the theorem actually applies to K, which is contrary to the assumption that $K \epsilon A$. We infer that A must be empty, proving the theorem.

Definition. Let H be a non-empty set of elements of F. H is called a *fractional ideal* if there exist an ideal K in D and an element a of F such that H consists of all numbers ak where $k \epsilon K$, $H = \{h | h = ak, k \epsilon K\}$. We write, briefly, $H = aK$.

The sum and difference of two elements of a fractional ideal H are again in H and if $h \epsilon H$ and $r \epsilon D$ then $rh \epsilon H$. (Why?) Any ideal K is also a fractional ideal as we see by putting $a = 1$ in the above definition. The multiplication of fractional ideals is defined just like the multiplication of "ordinary" ideals. Thus, if H_1 and H_2 are two fractional ideals, then H_1H_2 consists of all sums of products h_1h_2 where $h_1 \epsilon H_1$, $H_2 \epsilon H_2$. The product H_1H_2 is again a fractional ideal. If $H_1 = aK_1$, $H_2 = bK_2$, then $H_1H_2 = abK_1K_2$. Multiplication of fractional ideals is associative and commutative. (Why?) It reduces to the multiplication defined previously if the two fractional ideals happen to be ordinary ideals. While all elements of an ideal are integers (elements of D) a fractional ideal may contain also numbers that are not integers.

Let H be an ideal in D. We define H^* as the set of all elements h of F such that hk is an integer for all $k \in H$,

$$H^* = \{h | hk \in D \text{ for all } k \in H\} . \tag{1.1}$$

If $H = O$ then $H^* = F$. (Why?) If $H \neq O$ then we claim that H^* is a fractional ideal. H^* is certainly not empty since it contains D. Also, the sum and difference of the elements of H^* are elements of H^* and if $h \in K^*$ and $r \in D$ then $rh \in H^*$. (Why?) Now let $a \in H$ where $a \neq 0$. Such an a exists since we have assumed $H \neq O$. Let K be the set of all numbers ah where h varies over H^*,

$$K = \{k | k = ah,\ h \in H^*\} .$$

Then K is not empty and consists of integers. And if $k_1 \in K$ and $k_2 \in K$ then $k_1 = ah_1$, $k_2 = ah_2$ for certain $h_1 \in H^*$, and $h_2 \in H^*$. Hence $k_1 \pm k_2 = a(h_1 \pm h_2)$ where $h_1 \pm h_2 \in H^*$, and so $k_1 \pm k_2 \in K$. Finally, if $k \in K$, $k = ah$ where $h \in H^*$ and $r \in D$, then $rh \in H^*$, so $arh = rk \in K$. This shows that K is an ideal. But $H^* = (1/a)K$, and so H^* is a fractional ideal.

Theorem. If P is a prime ideal such that $P \neq D$ and P^* is defined as in (1.1) (with P for H) then P^* contains a number which is not an integer (that is, is not in D).

Proof. If $P = O$, so that P is the zero ideal, then $P^* = F$. In that case any number of F which is not an integer will serve, for example the number $\frac{1}{2}$. Suppose now that $P \neq O$, and let $a \in P$ such that $a \neq 0$. We consider the ideal (a). According to the first theorem of this section, there exist prime ideals $P_1, P_2, \ldots, P_m$, all different from 0, such that $P_1P_2 \cdots P_m \subset (a)$. Among such products we take one for which m is as small as possible. Suppose first that $m = 1$. Then $P_1 \subset (a)$ and, at the same time, $(a) \subset P$ since $a \in P$. Hence $P_1 \subset P$ and so (P_1 being prime) $P_1 = P$, which implies $P_1 = (a) = P$. (Why?) Since $P \neq D$ there exists a number b in D which does not belong to P, and hence, is not divisible by a. This means that the number b/a is not an integer. But $(b/a) \in P^*$, since any element of $P = (a)$ can be written as $h = ra$, $r \in D$, showing that $(b/a)h = (b/a)ra = rb$ is an integer. Accordingly, our assertion is proved when $m = 1$. Suppose now that $m > 1$. Since $(a) \subset P$ and $P_1P_2 \cdots P_m \subset (a)$, it follows that $P_1P_2 \cdots P_m \subset P$. Suppose that P does not contain P_1. According to our second definition of a prime ideal P must then contain $P_2P_3 \cdots P_m$, since it contains the product of P_1

and of $P_2P_3 \cdots P_m$. Suppose that P does not contain P_2 either. Then, by the same argument, P must contain the product $P_3 \cdots P_m$. Continuing in this way, we conclude that P contains at least one of the ideals $P_1, P_2, \ldots, P_m$, let us say $P_1 \subset P$, without loss of generality. But P_1 is prime, hence maximal, and $P \neq D$. We conclude that $P = P_1$, and so $PQ \subset (a)$ where $Q = P_2P_3 \cdots P_m$.

On the other hand Q cannot be contained in (a) since it is a product of $m - 1$ prime ideals, and no product of less than m prime ideals is contained in (a), by assumption. Let b be an element of Q which is not contained in a. It follows that b/a is not an integer. But b/a belongs to P^*, that is, $(b/a)h$ is an integer for any $h \in P$. This is true because the relation $PQ \subset (a)$ shows that $hb \in (a)$, that is, hb is divisible by a, that is, hb/a, which is the same as $(b/a)h$, is an integer. We have therefore established that b/a is in P^* although it is not an integer. This completes the proof of our theorem.

Next we prove

Theorem. Let P be a prime ideal, $P \neq O$, $P \neq D$, and let K be an ideal that is contained in P, $K \neq O$. Let P^* be defined by (1.1) (with P for H). Then $KP^* \neq K$.

Proof. Since $K \subset P$ it follows that $KP^* \subset PP^*$, and the elements of PP^* are all integers, $PP^* \subset D$. It follows that KP^* is a fractional ideal whose elements are integers, hence an ideal. (Why?) We have to show that $KP^* \neq K$.

Suppose that $KP^* = K$. Then $K(P^*)^2 = (KP^*)P^* = KP^* = K$, $K(P^*)^3 = (K(P^*)^2)P^* = KP^* = K$, and quite generally, $K(P^*)^n = K$ for $n = 1, 2, \ldots$. Now let β be an arbitrary element of P^*. Then $\beta^n \in (P^*)^n$, $n = 1, 2, \ldots$. Let $a \in K$ where $a \neq 0$. Then $a\beta^n$ belongs to $K(P^*)^n = K$ for $n = 1, 2, \ldots$. Hence, $a\beta^n = \mu_n$, where μ_n is an integer. In other words $\beta^n = \mu_n\gamma$ where $\gamma = 1/a$ and the μ_n are all integers. The result of Chapter VII, section 3, now forces us to conclude that β is an integer. But then all elements of P^* would be integers, contrary to the preceding theorem. Thus, the assumption that $KP^* = K$ leads to a contradiction, proving the assertion.

To conclude this section we shall show that, for $P \neq O$, P^* plays the role of the reciprocal, or inverse with respect to multiplication, relative to P. As we know, the position of the unit element, or neutral element with respect to multiplication for ideals, is occupied by D. Thus, we are going to prove

Theorem. Let P be a prime ideal, $P \neq O$. If P^* is defined by (1.1) (with P for H), then $PP^* = D$.

Proof. If $P = D$ then $P^* = D$ and $PP^* = D^2 = D$, which proves our assertion in this case.

Suppose now that $P \neq D$. Then PP^* is a fractional ideal whose elements are integers, hence an ideal. Also, $1 \in P^*$, so $P \subset PP^*$. Since P is prime we conclude that either $PP^* = P$ or $PP^* = D$. But $PP^* = P$ is ruled out by the preceding theorem (write P for K). Hence $PP^* = D$, as asserted.

2. PARADISE REGAINED

Unique Factorization Theorem for Ideals. Let K be any ideal in D other than O or D itself. Then there exist prime ideals P_1, $P_2, \ldots, P_m$ (where the same ideal may appear repeatedly), $m \geq 1$, $P_i \neq O$, $P_i \neq D$, $i = 1, 2, \ldots, m$ such that

$$K = P_1 P_2 \cdots P_m . \tag{2.1}$$

Also, if $Q_1, Q_2, \ldots, Q_l$ is any other sequence of prime ideals, $l \geq 1$, $Q_i \neq D$, $i = 1, 2, \ldots, l$, such that

$$K = Q_1 Q_2 \cdots Q_l , \tag{2.2}$$

then $m = l$, and, if we renumber the Q_i appropriately, then $P_i = Q_i$, $i = 1, 2, \ldots, m$.

To prove the first part of the theorem, let A be the set of all ideals K in D, $K \neq O$, $K \neq D$, for which there does not exist a sequence of prime ideals as described by the assertion. We have to show that A is empty. If A is not empty then, by the maximum principle, it contains a particular ideal, K, which is not contained in any other ideal of A. Now it follows from the first theorem of the preceding section that there exists a prime ideal, P, which includes A and which is different from D. For if all prime ideals P_i mentioned in that theorem were equal to D, we should have $P_1 P_2 \cdots P_m = DD \cdots D = D$, and this contradicts the fact that $P_1 P_2 \cdots P_m \subset K$ where $K \neq D$. Evidently $P \neq O$.

Consider the fractional ideal $H = KP^*$. Since $H \subset PP^* = D$, H contains only integers and is therefore an ideal. H contains K since $1 \in P^*$. But $H \neq K$, as shown by the last but one theorem of the preceding section. It therefore follows from the maximum property of K that H cannot belong to A. At the same time $H \neq D$ for if $H = D$, then $K = KD = K(P^*P) = (KP^*)P = DP = P$, so, in that case, K certainly satisfies the conclusion of the first part of the theorem and therefore cannot belong to A, contrary to assumption. But if $H \neq O$,

$H \neq D$ and $H \notin A$, then there exist prime ideals $P_1, P_2, \ldots, P_n$, all different from O and D such that $H = P_1P_2 \cdots P_n$. Hence

$$K = KD = K(P^*P) = (KP^*)P = HP = P_1P_2 \cdots P_nP .$$

This shows again that K satisfies the conclusion of the first part of the theorem and, accordingly, cannot belong to A. Thus, the assumption that A is not empty leads to a contradiction. This proves that any K, $K \neq O$, $K \neq D$, can be represented as a product of prime ideals as in (2.1), which is the assertion of the first part of the theorem.

The assertion of the second part of the theorem is that the representation of an ideal K other than O or D as a product of prime ideals (different from D) is unique except for the order of the factors. We shall prove this assertion by induction on m, beginning with $m = 1$. Suppose then that $K = P_1$, so that K is prime, and that, at the same time, $K = Q_1Q_2 \cdots Q_l$, $l \geq 1$ where the Q_i are prime, and P_1 and the Q_i are different from D. Evidently the Q_i are also different from O. Then $K = P_1 \subset Q_i$ and so $Q_i = P_1$, $i = 1, 2, \ldots, l$. If $l > 1$ we may then conclude that

$$\begin{aligned} D = P_1P_1^* = (Q_1Q_2 \cdots Q_l)P_1^* &= (Q_1 \cdots Q_{l-1})(P_1P_1^*) \\ &= (Q_1 \cdots Q_{l-1})D = Q_1 \cdots Q_{l-1} . \end{aligned}$$

But this implies that $D \subset Q_i$, $i = 1, \ldots, l-1$, and hence $Q_i = D$ for such i, contrary to assumption. Hence, $l = 1$ and $Q_1 = P_1$ proving our claim for $m = 1$.

Suppose now that we have proved the assertion for $m = n$. That is to say, suppose that we have shown that if $H = P_1 \cdots P_n = Q_1 \cdots Q_l$, $n \geq 1$, $l \geq 1$ and H and the P_i and Q_i are different from O and D, and the P_i and Q_i are prime, then $l = n$ and the Q_i may be renumbered so that $P_i = Q_i$, $i = 1, \ldots, n$. Let

$$K = P_1 \cdots P_nP_{n+1} = Q_1 \cdots Q_l ,$$

with the same assumptions on K and the P_i and Q_i. Then Q_l contains $K = P_1 \cdots P_nP_{n+1}$. Accordingly, by the second definition of a prime ideal, Q_l contains either P_{n+1} or $P_1 \cdots P_n$. If Q does not contain P_{n+1}, hence contains $P_1 \cdots P_n$, then it contains either P_n or $P_1 \cdots P_{n-1}$. Thus, by an argument employed earlier, we find that Q_l contains at least one of the P_i, and hence is equal to it. We may suppose without loss of generality that $Q = P_{n+1}$. Then

$$\begin{aligned} H = KP^*_{n+1} &= (P_1 \cdots P_n)(P_{n+1}P^*_{n+1}) = (P_1 \cdots P_n)D = P_1 \cdots P_n \\ &= (Q_1 \cdots Q_l)P^*_{n+1} = (Q_1 \cdots Q_{l-1})(P_{n+1}P^*_{n+1}) = Q_1 \cdots Q_{l-1} \end{aligned}$$

where $Q_1 \cdots Q_{l-1}$ has to be replaced by D if $l = 1$. Now since $n \geq 1$, $H \neq D$, as in the previous argument. At the same time, evidently, $H \neq O$. (Why?) Hence, $H = D$ is actually impossible, so that $l = 1$ is impossible. Thus $H = P_1 \cdots P_n = Q_1 \cdots Q_{l-1}$, $l - 1 \geq 1$. Hence, by what is supposed to have been proved already, $n = l - 1$, and a suitable renumbering of the Q_i yields $P_i = Q_i$ for $i = 1, 2, \ldots, n$. But this, together with $P_{n+1} = Q_l = Q_{n+1}$, proves the assertion also for $m = n + 1$, as required for our argument by induction. This completes the proof of the unique factorization theorem for ideals.

3. FACTORIZATION OF IDEALS IN SPECIAL CASES

Suppose first that $D = J$, the ring of rational integers, and $F = Ra$, so that all ideals in D are principal ideals. Consider any ideal $K = (a)$ in D, where $K \neq O$ and $K \neq D$. This will be the case if a is neither zero nor a unit, that is, precisely when $a \neq 0$, $a \neq 1$, $a \neq -1$. Let $K = P_1P_2 \cdots P_m$, $m \geq 1$, where the P_i are prime ideals other than O or D. If $P_i = (p_i)$, where we take p_i to be positive, then p_i is prime. Thus $K = P_1P_2 \cdots P_m$ may be rewritten as

$$(a) = (p_1)(p_2) \cdots (p_m) = (p_1p_2 \cdots p_m)\,.$$

But this shows that $a \sim p_1p_2 \cdots p_m$, in other words

$$a = \pm p_1p_2 \cdots p_m\,,$$

which is the representation of a as a product of prime numbers, in the usual way. At the same time the *uniqueness* of the factorization into prime ideals corresponds to the fact that if $p_1p_2 \cdots p_m = q_1q_2 \cdots q_l$, $m \geq 1$, $l \geq 1$, where the p_i and q_i are prime, then $l = m$ and we may renumber the q_i so that $p_i \sim q_i$. Indeed, $p_i \sim q_i$ is equivalent to $(p_i) = (q_i)$. We may therefore say that even in the simplest case, that is, when $D = J$, the factoring of (a) into prime ideals is to some extent "more unique" then the factoring of the number a into primes, since the ideals (p_i) and (q_i) are equal even if the numbers p_i and q_i are only associated. Similar remarks hold for rings $J(\sqrt{d})$ in which every ideal is principal.

However, the factorization of an ideal into prime ideals gains real interest for rings $J(\sqrt{d})$ in which the factorization into prime numbers is no longer unique even in the weaker sense of counting associated numbers as "roughly the same." We have seen that $J(\sqrt{10})$ is such a ring. More particularly, we proved that we can write the number 6 in two different ways as a product of prime numbers in $J(\sqrt{10})$,

$6 = 2 \cdot 3 = (4 + \sqrt{10})(4 - \sqrt{10})$ where neither $2 \sim 4 + \sqrt{10}$, nor $2 \sim 4 - \sqrt{10}$, and neither $3 \sim 4 - \sqrt{10}$ nor $3 \sim 4 + \sqrt{10}$. We shall now show how to write the ideal (6) as a product of prime ideals (different from $J(\sqrt{10})$) where we know that, except for the order in which the factors are written, this can be done only in one way.

We define three ideals in $J(\sqrt{10})$, P_1, P_2, P_3, by

$$P_1 = (2, \sqrt{10}), \quad P_2 = (3, 4 + \sqrt{10}), \quad P_3 = (3, 4 - \sqrt{10}) .$$

Then

$$P_1^2 = (2, \sqrt{10})(2, \sqrt{10}) = (4, 2\sqrt{10}, 10) . \tag{3.1}$$

We claim that $P_1^2 = (2)$. Indeed $P_1^2 \subset (2)$, since all the elements of the basis of P_1^2 which is written on the right-hand side of (3.1), that is, $4, 2\sqrt{10}, 10$, are divisible by 2. Thus, in order to prove that $P_1^2 = (2)$ we only have to show that the number 2 is included in P_1^2. But $2 = 10 - 2 \cdot 4$, and 10 and 4 are elements of the basis of P_1^2 under consideration. Hence, $2 \in P_1^2$, $P_1^2 = (2)$.

Next, we shall show that $P_2P_3 = (3)$. Indeed

(3.2)

$$P_2P_3 = (3, 4 + \sqrt{10})(3, 4 - \sqrt{10}) = (9, 12 - 3\sqrt{10}, 12 + 3\sqrt{10}, 6)$$

All elements of the basis of P_2P_3 on the right-hand side of (3.2) are divisible by 3, so that $P_2P_3 \subset (3)$. On the other hand, 3 is contained in P_2P_3 since $3 = 9 - 6$, that is, 3 equals the difference between the first and fourth elements of the basis. Thus, $3 \in P_2P_3$, $P_2P_3 = (3)$ and $P_1^2P_2P_3 = (2)(3) = (6)$.

We shall now show that the ideals P_1, P_2, P_3 are prime. Since $P_1 = (2, \sqrt{10})$ it is evident that any number $\beta = 2a + \sqrt{10}b$ where a and b are *rational* integers, is included in P_1. Conversely, let β be any number which is included in P_1. Since 2 and $\sqrt{10}$ form a basis for P_1 it must be possible to write β in the form $\beta = 2\gamma_1 + \sqrt{10}\gamma_2$, where γ_1 and γ_2 are elements of $J(\sqrt{10})$ which are not necessarily rational Let $\gamma_1 = c_1 + \sqrt{10}d_1$, $\gamma_2 = c_2 + \sqrt{10}d_2$, where c_1, d_1, c_2, d_2 are rational integers. Substituting these expressions for γ_1, γ_2 in $\beta = 2\gamma_1 + \sqrt{10}\gamma_2$, we obtain

$$\beta = 2(c_1 + \sqrt{10}\, d_1) + \sqrt{10}(c_2 + \sqrt{10}\, d_2) =$$
$$2c_1 + 10d_2 + \sqrt{10}(2d_1 + c_2) .$$

This shows that $2c_1 + 10d_2$ is an even rational integer (it is divisible by 2). We conclude that the elements of P_1 are *precisely* the numbers of

$J(\sqrt{10})$ that can be written in the form $\beta = 2a + \sqrt{10}\, b$ where a and b are arbitrary rational integers. Let $\beta_1 = a_1 + \sqrt{10}\ b_1$ and $\beta_2 = a_2 + \sqrt{10}\ b_2$, then $\beta_1\beta_2 = (a_1a_2 + 10b_1b_2) + \sqrt{10}(a_1b_2 + b_1a_2)$. In order to establish that P_1 is prime we have to verify that if neither β_1 nor β_2 belong to P_1, then their product, $\beta_1\beta_2$, does not belong to P_1 either. But if β_1 and β_2 do not belong to P_1 then a_1 and a_2 are odd, and so a_1a_2 is odd and $a_1a_2 + 10b_1b_2$ is odd. This shows that $\beta_1\beta_2 = (a_1a_2 + 10b_1b_2) + \sqrt{10}(a_1b_2 + b_1a_2)$ cannot belong to P_1, and proves that P_1 is prime.

Consider the ideal $P_3 = (3,\ 4 - \sqrt{10})$, and let β be an arbitrary element of $J(\sqrt{10}), \beta = a + \sqrt{10}\, b$, where a and b are rational integers. We are going to show that β belongs to P_3 if and only if the sum $a + b$ is divisible by 3.

Suppose first that $a + b$ is divisible by 3, $a + b = 3k$, where k is a rational integer. Then $a = 3k - b$ and so

$$\beta = a + \sqrt{10}\, b = 3k - b + \sqrt{10}\, b = 3(k + b) - (4 - \sqrt{10})b\,.$$

But any number which can be expressed in the form $3\gamma_1 + (4 - \sqrt{10})\gamma_2$ where γ_1 and γ_2 are integers in $J(\sqrt{10})$ belongs to P_3. Accordingly, $\beta \in P_3$.

Suppose next that β belongs to P_3. Then $\beta = 3\gamma_1 + (4 - \sqrt{10})\gamma_2$ where $\gamma_1 = a_1 + \sqrt{10}b_1$, $\gamma_2 = a_2 + \sqrt{10}b_2$ for certain rational integers a_1, b_1, a_2, b_2. Hence

$$\begin{aligned}\beta &= 3(a_1 + \sqrt{10}\, b_1) + (4 - \sqrt{10})(a_1 + \sqrt{10}\, b_2)\\ &= 3a_1 + 4a_2 - 10b_2 + \sqrt{10}(3b_1 - a_2 + 4b_2)\,.\end{aligned}$$

Comparing this with the expression $\beta = a + \sqrt{10}\ b$, we find that $a = 3a_1 + 4a_2 - 10b_2$, $b = 3b_1 - a_2 + 4b_2$, and so $a + b = 3a_1 + 3b_1 + 3b_1 + 3a_2 - 6b_2$. Hence $a + b = 3(a_1 + b_1 + a_2 - 2b_2)$, $a + b$ is divisible by 3 as asserted.

Now consider the product $\beta_1\beta_2 = (a_1a_2 + 10b_1b_2) + (\sqrt{10})(a_1b_2 + b_1a_2)$ of the numbers $\beta_1 = a_1 + \sqrt{10}\, b_1, \beta_2 = a_2 + \sqrt{10}\, b_2$, where a_1, b_1, a_2, b_2 are again rational integers. Putting $\beta_1\beta_2 = a + \sqrt{10}\, b$ where a and b are rational integers, we see that $a = a_1a_2 + 10b_1b_2$, $b = a_1b_2 + b_1a_2$, $a + b = a_1a_2 + 10b_1b_2 + a_1b_2 + b_1a_2 = a_1a_2 + a_1b_2 + b_1a_2 + b_1b_2 + 9b_1b_2 = (a_1 + b_1)(a_2 + b_2) + 9b_1b_2$. It follows that $a + b$ is divisible by 3 if and only if $(a_1 + b_1)(a_2 + b_2)$ is divisible by 3. (Why?) If neither β_1 nor β_2 belongs to P_3, then neither $a_1 + b_1$ nor $a_2 + b_2$ is divisible by 3. Therefore $(a_1 + b_1)(a_2 + b_2)$ and hence $a + b$ are not divisible by 3 either, which implies that $\beta_1\beta_2$ does not belong to P_3. We conclude that P_3 is prime.

The proof that P_2 is prime is quite similar. Accordingly, the formula $(6) = P_1^2P_2P_3$ represents the factorization of the ideal (6) into prime ideals.

4. CONCLUDING REMARKS

We have come to the end of our road. If you are interested in going on you may find out that many of our results can be generalized and deepened with relatively little effort. Thus, it is natural to consider not only quadratic fields but fields which contain all expressions $a_0 + a_1\theta + a_2\theta^2 + \cdots + a_n\theta^n$ where the a_i take arbitrary rational values and θ is a fixed number which is a root of an irreducible polynomial of degree n. And you may consider in particular the integers of such a field, that is, those numbers which are roots of *monic* polynomials with rational integer coefficients. These form an integral domain within which the theory of ideals can be developed in much the same way as for quadratic fields. Beyond that the theory of ideals has many more applications. Here we shall be content if we have managed to show you some of the beauty of the subject in a concrete case.

Suggestions for Further Reading

1. LeVeque, W. J., *Elementary Theory of Numbers*, Addison-Wesley, Publishing Company, Reading, Mass., 1962.
2. McCoy, W. H., *Rings and Ideals*, Carus Mathematical Monographs, No. 8, Mathematical Association of America, 1948.
3. Hardy, G. H., and E. M. Wright, *An Introduction to the Theory of Numbers*, Oxford University Press, Oxford, 4th ed., 1960.
4. Niven, I., and H. S. Zuckerman, *An Introduction to the Theory of Numbers*, John Wiley & Sons, New York, 1960.
5. Pollard, H., *The Theory of Algebraic Numbers*, Carus Mathematical Monographs, No. 9, Mathematical Association of America, 1950.
6. Weiss, E., *Algebraic Number Theory*, McGraw-Hill Book Company, New York, 1963. (This book requires a good deal of preparation in contemporary abstract terminology).

Index

This book is set in Linotype Modern No. 21, with formulas in Monotype Modern No. 8, headings in Caledonia bold, and chapter titles in Craw Clarendon Book. The composition, printing, and binding are by Kingsport Press. The book is printed by letterpress on Perkins & Squier Antique paper, and bound in Columbia Riverside vellum.